D0712159

BASIC MATHEMATICS

An Introduction

BASIC MATHEMATICS
An Introduction

Alan Graham

First published in Great Britain in 1995 by Hodder Education. An Hachette UK company.

First published in US in 1997 by The McGraw-Hill Companies, Inc.

This edition published 2017 by John Murray Learning

Previously published as *Teach Yourself Basic Mathematics*, *Basic Mathematics: Teach Yourself* and *Mathematics: A Basic Introduction*.

Copyright © 1995, 2001, 2003, 2008, 2010, 2013, 2017 Alan Graham

British Library Cataloguing in Publication Data: a catalogue record for this title is available from the British Library.

Library of Congress Catalog Card Number: on file.

ISBN 9781473651975

eISBN 9781473651982

Typeset by Cenveo® Publisher Services.

Printed and bound in Great Britain by CPI Group (UK) Ltd., Croydon, CR0 4YY.

John Murray Learning policy is to use papers that are natural, renewable and recyclable products and made from wood grown in sustainable forests. The logging and manufacturing processes are expected to conform to the environmental regulations of the country of origin.

Carmelite House

50 Victoria Embankment

London EC4Y 0DZ

www.teachyourself.com

Contents

Meet the author

Welcome to *Basic Mathematics*!

If you found that maths was something of a closed book at school, please don't feel intimidated or frightened by the prospect of opening up and reading through this book now. Having spent most of my adult life teaching adults at the Open University, I have come to realize that 90% of people's problems with the subject stem from fear – fear of failure, fear of showing yourself up in front of others, fear of feeling lost and confused. All of these negative feelings combine to bring the shutters down on your learning to the point where you are no longer able to listen or even think.

You may be wondering what has changed to enable you to succeed now where you didn't before. The answer is, everything! You may not fully realize it, but as a functioning adult who has survived in the world beyond school, you have actually picked up a lot of useful skills on the way:

▶ You have a much better understanding of *how the world works* – something that will be invaluable when it comes to understanding, selecting and using mathematical skills.

▶ You have greatly developed your *thinking skills* in the course of tackling the many decisions and problems of your everyday life.

▶ Most importantly, you almost certainly now have a *confidence* and a *desire to learn* that you didn't possess at the age of 15.

All of these factors will combine to raise the shutters of your mind, so that it is truly open to the clear thinking needed to understand and master mathematical skills.

And finally, remember that there is nothing that succeeds like success. The more you find yourself getting the exercises right, the more you'll want to tackle the next one, and gradually you will really start to believe that, yes, I'm becoming a mathematician!

Acknowledgements

Many thanks to Wendy Austen, Carrie Graham, James Griffin and Sally Kenny for their help in the preparation of this book.

Introduction

Mathematics is about numbers. But, more than that, it is about describing and explaining *patterns* – that includes number patterns (*arithmetic*) but also patterns in shape and space (*geometry*). Increasingly, our 'digital' world bombards us with numerical information and an essential life skill is being able to generate, collect, observe and explain patterns when handling data (*statistics*). Finally, mathematics is about responding to an inner human need to pose and solve problems, be they everyday problems or those of the inventively curious mind.

There are four main strands of basic mathematics:

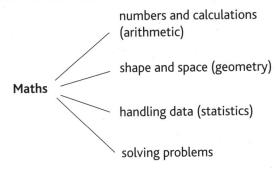

Maths
- numbers and calculations (arithmetic)
- shape and space (geometry)
- handling data (statistics)
- solving problems

In the main part of the book you will cover these four themes; you will also learn what algebra is about and discover how to use a spreadsheet.

You will have the chance to try a variety of puzzles such as the one below:

> What are the next two letters in this sequence:
>
> O, T, T, F, F, S, S, E, ..., ... ?

Finally, at the end of this book, you will find 14 case studies showing you how mathematics can help to solve everyday problems and provide insights that would not otherwise have been obvious.

The next two letters in this sequence are N and T (the letters are the initials of 'One', 'Two', 'Three', ...

Numbers and calculations

For most people, the topics *numbers* and *calculations* represent the core of the mathematics that they learnt at school; some basic ideas about numbers are explained in Chapters 2 and 3. Clearly a sound understanding of basic maths is useful, not just for our own personal needs, but also to be able to help others (your child, for example). So, these two chapters will help you to present ideas about numbers to others and to clarify them for yourself. Chapter 2 will also help you to become more familiar with a simple four-function calculator.

Fractions, decimals and percentages

Moving beyond whole numbers raises the question of what lies in-between them. These 'broken bits' of numbers can be written using fractions, decimals and percentages. These each require their own special notation and include numbers like 'three point seven', 'four and three quarters', 'twenty-five per cent', and so on. In Chapters 4, 5 and 6, you will get a straightforward explanation of what fractions, decimals and percentages are, how they are written (e.g. 3.7, $4\frac{3}{4}$ and 25%, respectively), why they are written that way and how to perform calculations with them.

Measuring

How many times would you say you perform some measurement task during the course of an average day? The answer is probably much more often than you think. Sometimes the measurement you do is *formal* – perhaps reading the time on a watch or clock, checking the temperature on a thermostat, reading the dial (in litres or in money) as you fill the car up with petrol, checking your weight on the bathroom scales and so on.

There may be even more occasions where you are engaging in *informal* measurement and this often involves making judgements based on using estimation skills – like working out whether the armchair will fit into the gap between the sideboard and the door, or choosing how many potatoes to cook for a family of four. Informal measurement often involves the use of some everyday template or benchmark against which to make comparisons. For example, when estimating, say, 100 metres, I tend to think about the length of a football pitch or when estimating weights I think of a bag of sugar (1 kilo).

It is worth remembering that numbers which arise in the everyday world have had to come from somewhere and the chances are that they are the result of some measuring process. But in order to handle these numbers sensibly and with understanding, you need to know how they were measured and in which measuring units. These are some of the issues explored in Chapter 7.

Statistical graphs and charts

It is claimed that about 80% of all the data in the world today has been created in the past two years. According to computer company IBM, 2.5 exabytes – that is 2.5 billion gigabytes – of data was generated *every day* in 2012. This is not a static figure – estimates suggest that it is growing by roughly 30% each year. It is hard to avoid the conclusion that we are drowning in statistical data. And unfortunately, for most people, the sight of a page full of figures gives them the heeby-jeebies.

There are two main ways of seeking out helpful patterns in data. One is to use summaries, like reducing the many numbers to a simple average. The other is to depict the data visually, using a statistical chart. It is the second approach that is explored here. A number of such charts are explained in Chapter 8, including barcharts, piecharts, scattergraphs and line graphs. You will also learn some of the tricks and dodges used by advertisers and others to deliberately mislead the gullible public. You have been warned!

Algebra

And then there's algebra. I have often heard students observe when thinking back to their school days, that, just when they were getting to grips with doing calculations involving numbers, along came letters like x and y, and a and b to really fox them. So, what is the point of doing calculations with letters? For most of us, school algebra was just about learning the rules of handling expressions and equations involving letters, but we never actually got to discover why that might be a useful thing to do. This is a bit like a musician practising scales and arpeggios but never actually getting to play a tune. It is a question that you should get a proper answer to in Chapter 9.

Modelling with mathematics

Mathematics is usually defined in terms of the sets of skills it uses – arithmetic, algebra, geometry, and so on. If your mathematics is to be useful to you in the real world, knowing how to perform these skills is clearly important. But just as important is the ability to know what skills to use in any given situation, and also being able to translate your mathematical answer back into the real world situation where the problem originated.

For your mathematical skills to be useful in solving everyday problems, there are three stages involved: deciding what 'sum' to do, doing the 'sum' and interpreting the result. This is often referred to as *mathematical modelling*. In the Appendices, at the end of the book, you are presented with 14 everyday scenarios: a question is posed, and you are taken through the solutions step by step. Remember that successfully tackling these sorts of real problems is just as much about using thinking skills and strategies, as it is about being able to perform the calculations.

How to succeed at maths

Finally, here are four useful tips to help you to make your maths journey an exciting and successful experience:

▶ Go at your own speed and be prepared to re-read sections if necessary.

▶ Where possible, relate the ideas to your own personal experiences.

▶ Try to learn actively (this means actually doing the exercises, rather than saying vaguely, 'Oh yes, I'm sure I can do that').

▶ Be prepared to explore and experiment with ideas, particularly by taking every opportunity to exploit technologies like the calculator and computer.

1

Reasons to be cheerful about mathematics

In this chapter you will learn:

▶ *why you can succeed at learning maths, even if you have failed in the past*

▶ *why a calculator will help*

▶ *how to develop a positive attitude to maths.*

Is this book really for me?

I wonder what made you decide to buy this book! No doubt each individual has his or her own special reason. As author, I clearly can't meet everyone's exact needs, so I tried to write this book with two particular types of reader in mind. I will call them Marti and Mel.

MARTI

Marti was neither particularly good nor particularly bad at mathematics, but was able to get by. She could usually work out the sums set by her teacher at school, but never really understood why they worked. She had a vague sense that there was, potentially, an exciting mental world of mathematical ideas to be explored, but somehow it never happened for her. She now has two children aged ten and six. She is aware that her own attitudes to mathematics are being passed on all the time to her children, and she wants to be able to encourage and help them more effectively. The crunch for Marti came when her daughter asked her about the difference between odd and even numbers. Marti knew which numbers were odd and which were even, but she couldn't really explain why. She might have bought this book to understand some of the basic 'whys' in mathematics.

MEL

Mel never got on with mathematics in school. He lost confidence with it at an early stage and constantly had the feeling of 'if only the teacher and the other pupils knew how little I know, they'd be shocked'. As the years went on, he learned to cover up his problems, and as a result he always had a bad feeling whenever mathematics cropped up – a combination of fear of being caught out and guilt at not properly facing up to it. He has a good job, but his fear of mathematics regularly causes him problems. He might have bought this book in order to lay to rest the ghost of his mathematical failure once and for all.

While your name is unlikely to be Mel or Marti, maybe there is an aspect of their experiences of, and hopes for, mathematics learning that you can identify with?

SHOULD I START AT CHAPTER 1 AND READ RIGHT THROUGH THE BOOK?

The honest answer to this question is, 'It all depends…'. If your mathematics is reasonably sound, you might like to skip Chapters 1 to 3, which deal with basic ideas of what a number is (hundreds, tens and units), how to add, subtract, multiply and divide and how to make a start with a simple four-function calculator. However, even if you do have a basic understanding of these things, you could always skim-read these chapters, if only to boost your confidence.

If, like Marti, you have an interest in helping someone else, say a child, then read these three chapters with a teacher's hat on. They should provide you with some ideas about how you can help someone else to grasp these ideas of basic arithmetic.

Can I succeed now if I failed at school?

There are a number of reasons that school mathematics may have been dull and hard to grasp. Here are some resources that you may have now which weren't available to you when you were aged 15.

RELEVANCE

There were undoubtedly many more important things going on in most people's lives at the age of 8, 12 or 15 than learning about adding fractions or solving equations. Mathematics just doesn't seem to be particularly important or relevant at school. But as you get older, you are better able to appreciate some of the practical applications of mathematics, and how various mathematical ideas relate to the world of home and work. Even a willingness to entertain abstract ideas for their curiosity value alone may seem more attractive out of a school context. One student commented:

> Yes, I think you can associate more with it. When you get to our age, when you've a family and a home, I think you can associate more if you do put it to more practical things. You can see it better in your mind's eye.

CONFIDENCE

As an adult learner, you can take a mature approach to your study of mathematics and be honest in admitting when you don't understand. As one adult learner said, after an unexpectedly exciting and successful mathematics lesson conducted in a small informal setting:

I felt I could ask the sort of question that I wouldn't have dared ask at school – like 'What is a decimal?'

Spotlight

For many years I have been lucky enough to work for the Open University – sometimes referred to as the 'second chance' university. Many of our students, through no fault of their own, got off to a bad start learning maths at school and disliked the subject. How different it is now. Watching these students at summer school I see excitement, enthusiasm, confidence and – for the first time for many of them – the experience of success in a subject they thought they couldn't do!

MOTIVATION

A third factor in your favour now is motivation. Children are *required* to attend school and to turn up to their mathematics lessons, whether they want to or not. In contrast, you have made a conscious choice to study this book, and the difference in motivation is crucial. Perhaps you have chosen to read the book because you need a better grasp of mathematics in order to be more effective at work. Or maybe you are a mathephobic parent who wants a better outlook for your own child. Or it is possible that you have always regretted that your mathematical understanding got lost somewhere, and it is simply time to lay that ghost to rest.

MATURITY AND EXPERIENCE

You are a very different person from when you were 8 or 15 years old. You now have a richer vocabulary and a wide experience of life, both of which will help you to grasp concepts you never understood before.

Hasn't maths changed since I was at school?

Mathematics *has* changed a bit since you were at school, but not by as much as you think. The changes have occurred more in the language of mathematics than in the topics covered. Basic arithmetic is still central to primary mathematics, and today's 12-year-olds are still having the same sorts of problems with decimals and fractions as you did.

Spotlight

Visiting schools is a part of my job. In an infants' classroom I was enjoying watching these six- and seven-year-olds excitedly using calculators for the first time. 'What's this little dash, Miss?' asked one girl. She had just tried subtracting a bigger number from a smaller number on her calculator and had discovered the world of negative numbers all by herself.

Will it help to use a calculator?

Yes. Most adults have a calculator, but perhaps many rarely use them. This book should help you overcome any anxiety you may feel. As one student explains:

It's only a thing with buttons, isn't it? All right, it takes you a while to know what each button is, but it's like driving a car. Once you've learnt which key to go to, you do it and that's it. It's just a case of learning. No, it didn't frighten me.

You will need to get a calculator before reading the following chapters. You won't need anything more sophisticated (or more expensive) than a simple 'four function' calculator (i.e. one which does the four functions of add, subtract, multiply and divide).

There are a number of reasons why this book has been written assuming a calculator is to hand, and some of these are spelled out in more detail at the start of the next chapter. Two reasons are as follows. Firstly, the calculator is more than simply a calculating device. As you work through this book, you will see how the calculator can be used as a means of learning and exploring mathematics. As one student commented:

> *That was something that I enjoyed; knowing that you can divide and get larger numbers. I mean, that was a thing that, without the calculator, I would have been, ... would never have believed, but because it was instant, it was something I could see straightaway.*

Secondly, being competent with mathematics isn't just some abstract skill that divides those who can from those who can't. Mathematical skills should be useful and insightful to you in your real world and in the real world that most of us inhabit.

This chapter ends with a section on children's mathematics. This theme will recur throughout the book, and I hope that even if you are not yourself a parent, you will find it interesting. My experience of working with adults has been that looking at children's errors and experiences in mathematics can be a fascinating 'way in' for the adults, giving a better understanding of the key ideas of mathematics.

Can I help my child?

Many parents who were unsuccessful at mathematics themselves have the depressing experience of watching their children follow in mother's or father's footsteps. Being hopeless at mathematics seems to run in families – or does it? While it certainly seems likely that *confidence* in mathematics passes from parent to child, it is less certain that 'mathematical ability' is inherited in this way. So, if you consider yourself to be a duffer at mathematics, there is no reason to condemn your child to the same fate. Indeed, there is much that you can do to build your child's confidence and stimulate his or her interest in the subject.

Try to stimulate your child's thinking and curiosity about mathematical ideas. The ideal setting for these conversations might be in the kitchen, in a supermarket or on a long journey.

How, then, can parents encourage in their children a curiosity and excitement about mathematical ideas? Below are a few general pointers to indicate the sort of things you might say and do with your child to achieve this aim, by creating what we professionals helpfully call a 'mathematically stimulating environment':

▶ Where possible try to respond to your child's questions positively.

▶ Encourage the desire to have a go at an answer, even if the answer is wrong. Wrong answers should be opportunities for learning, not occasions for punishment or humiliation.

▶ Try to think about what makes children tick and the sort of things they might be curious about.

▶ Resist providing easy adult answers to your child's questions, but rather try to draw further questions and theories from her/him.

▶ If a child gives a wrong answer it is probably for a good reason. Try to discover the cause of the problem – it could be, for example, that your question was not clearly expressed, or that the child is simply not ready for the concept.

▶ There may be specific educational toys and apparatus which are helpful to have around. Having said that, a four-year-old will learn to count as effectively or ineffectively with pebbles or bottle tops, as with 'proper' counting bricks!

▶ Finally, remember that 'how' and 'why' questions are at a higher level of curiosity than 'what' questions, and these should be encouraged. (For example, '*why* does $2 \times 3 = 3 \times 2$?' is a more stimulating question than 'what is 2×3?')

The biggest barrier to learning mathematics is *fear*: the fear of being shown up as not understanding something which seems to be patently obvious to the rest of humanity. The best way of helping your child is to start by trying to overcome that fear in yourself. It doesn't matter if you don't know a simple fraction from a compound fracture; your competence at mathematics is less important than your willingness to be honest about what you don't understand yourself.

Just as with cooking, carpentry or playing the piano, learning mathematics doesn't happen just by reading a book about it. Although explanations can lead to understanding, practice is necessary if you are to achieve mastery.

Spotlight

I once asked my son if his mathematical understanding benefited from having a maths teacher as a dad. He said, 'Yes, but not to get long detailed explanations of things. I just wanted short answers to simple things where I was stuck. Then I could go away and work the rest out for myself.'

The magic number machine

In this chapter you will learn:

▶ *how to say 'hello' to your calculator*
▶ *about numbers and how they are represented*
▶ *how children first find out about numbers.*

Note: In this and all subsequent chapters, comments on exercises are given at the end of the chapter.

Numbers, numbers everywhere

Everywhere you look, numbers seem to leap out. They lie hidden in recipe books, are stamped onto coins and printed onto stamps, flash up on supermarket check-outs, provide us with breakfast reading on cornflakes packets, are displayed on buses, spin round on petrol pump dials ... It seems that, whatever task we want to perform, numbers have some role to play. Here are a few more examples.

▶ A farmer will check that all the cows are in by counting them as they go through the gate.

▶ At church, the vicar reads out the hymn number. Members of the congregation are able to find the hymn in their hymn books only if they know how numbers are organized in sequence.

▶ The recipe book says 'pour into a 9-inch square tin and bake at 180°C for 45–50 minutes'. For this to make sense, we need to know about measurement of size in inches, temperature in degrees Celsius and time in minutes.

▶ Competitive sport is totally based on numbers, usually in the form of *scoring* – half-time scores, number of runs, highest break, winning times ...

You can probably think of lots more examples of your own. It's hard to imagine a world with no numbers. How would we survive without them? What alternative ways might we think up to organize human activity if we couldn't count on numbers to control them?

Exercise 2.1 is designed to help you to be more aware of the role that numbers play in your life.

Exercise 2.1 How long can you last without numbers?

Take a waking period of, say, 30 minutes of your life, and see how many situations require some use and understanding of numbers. Then read on.

I tried the exercise while writing this chapter on my computer. Within a few seconds I found that I was increasing the viewing Zoom on the document to 150%.

Minutes later the phone rang and I was aware that the person on the other end had just pressed a series of *numbers*.

The phone call was to fix a meeting, so I had to confirm the *date* and *time* in my diary.

Shortly afterwards, I went to make a cup of coffee (I don't have a very long attention span!). This involved putting milk and water into a cup, placing it into the microwave and keying 1 45 (1 minute and 45 seconds) into the *timer*.

Numbers are the building blocks of mathematics. So, just like being able to read, there are certain basic mathematical skills that you need in order to live a normal life. For example, being able to:

▶ read numbers and count
▶ tell the time
▶ handle money when shopping
▶ weigh and measure
▶ understand timetables and simple graphs.

Spotlight

The word *number* crops up in many phrases and sayings. To 'do a number' on someone means to cheat them. When your 'number is up', then you'd better watch out because you may be about to die. If you 'have someone's number', you have worked out what sort of person they are (and it's not good!). The phrase 'painting by numbers' means doing something that should be creative in a mechanical and unfeeling way.

As well as having a useful, practical side, solving problems with mathematics can also be challenging and fun. Anyone who has done dressmaking or carpentry knows that mathematics can be used in two ways. One is in the practical sense of measuring and using patterns and diagrams. The other is more abstract – 'How can I cut out my pattern so as to use up the least material?', 'Can I use the symmetry of the garment to make two cuttings in one?' The pleasure you get from solving these sorts of problems is what has kept mathematicians going for four thousand years!

Most of all, mathematics is a powerful tool for expressing and communicating ideas. Sadly, too few people ever get a sense of the 'power to explain' that mathematics offers.

Let us return now to these building blocks (the numbers) and see what sense your calculator makes of them.

Saying hello to your calculator

Most basic calculators look something like this:

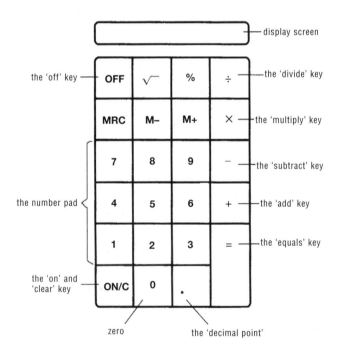

Before we tackle the hard stuff, you should start by saying hello to your calculator. Usual etiquette is as follows:

▶ press the 'On' switch (probably marked $\boxed{\text{ON}}$ or $\boxed{\text{ON/C}}$)

▶ now key in the number 0.7734

▶ turn your calculator upside down and read its response on the screen.

Good! Having now exchanged 'hELLOs', this is clearly the basis for a good working relationship!

As was explained in the previous chapter, this book aims to teach you basic mathematics *with a calculator*. There are a number of good reasons for presenting the book in this way, some of which are listed below.

▶ For some types of problem, calculators take the slog out of the arithmetic and allow you to focus your attention on understanding what the problem is about.

▶ Examination boards have largely designed their examinations on the assumption that all candidates have access to a calculator.

▶ Most importantly, the calculator provides a powerful aid to learning mathematics. See if you are more convinced about this by the time you have finished the book.

Spotlight

The word 'minus' is a source of great confusion in arithmetic, as it has two possible meanings. One meaning is 'subtract' (as in: 6 *minus* 2 equals 4) and the other is as a negative number (as in: the number that is 3 less than zero is *minus* 3). If your calculator has just one minus key (as does the calculator above), then you can use it for both purposes – to subtract one number from another, and to enter a negative number. More sophisticated calculators provide two different 'minus' keys for these two rather different functions.

Introducing the counting numbers

Let us begin at the beginning, with the counting numbers
1, 2, 3, 4, 5 …

No doubt they look familiar enough. But before you reach for
pencil and paper, let's see how they look on your calculator.

Exercise 2.2 Entering numbers

Switch on your calculator and key in the following sequence

1 2 3 4 5 6 7 8 9

Look carefully at the screen.

Write down what is recorded and make a note of exactly how the
numbers are displayed.

If you look carefully at each number, you should see that it is
made up of a series of little dashes. For example, the 3 requires
five dashes and is written as follows.

You might like to consider in the next exercise which of the
numbers from 0 to 9 requires the fewest and which the most
dashes to be displayed.

Exercise 2.3 Displaying numbers

a Check how the other numbers are represented and count the
number of dashes each number needs. Then put your answer
into the table below (one, the 3, has already been done for you).

Number	0	1	2	3	4	5	6	7	8	9
Dashes				5						

b See if you can sketch any new combinations of dashes that are
not already covered by the numbers. If you were a calculator
designer, could you turn these into anything useful?

Your table should now help you to answer the question posed earlier, namely to spot which number requires the fewest and which the most dashes to be displayed. Was your earlier guess correct? In fact, the number 1 uses the fewest dashes (two), while the 8 uses the most (all seven).

There are a few combinations not covered by the numbers. For example:

and

One possible application of these shapes is to spell letters. For example, the 'E' above could be used to refer to 'Error' if an impossible key sequence was pressed. The 'C' could perhaps be used to mean 'Constant operating' or maybe 'Careful, you're hitting my keys too hard!'

The calculator constant

Most calculators can be set up to produce number sequences. Here is a simple one to try.

> Press ⊞ 1
> Now press ⊟ repeatedly.

You should see the counting numbers in sequence: 2, 3, 4, 5, 6 ... If this has not happened, try the following.

> Either: Press 1 ⊞ ⊞ and then the ⊟ repeatedly.
> Or: Press 1 ⊞ 1 and then the ⊟ repeatedly.
> Or: Read your calculator manual.

What is going on here is that your calculator is doing a constant 'add 1' calculation.

Spotlight

'Constant' calculations are available on most calculators. How they can be used will be explained later.

Tens and units

Let's have another look at the sequence of counting numbers.

Repeat the instruction above, using the constant to add 1.
As before, repeatedly press ⌐=⌐ until the calculator displays 9.
The display should look like this.

| 9 |

Pause for a moment and then press the ⌐=⌐ once more.

You should now see the following.

| 10 |

Most people will recognize that this is simply a ten, but note
that it is quite different from the previous numbers. Focus on
the fact that there are now not one but two figures displayed
here. What has happened in the move from nine to ten is that
the nine has changed to a zero and a new figure, the 1, has
appeared to the left of the zero, thus:

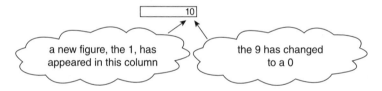

a new figure, the 1, has
appeared in this column

the 9 has changed
to a 0

It is worth reflecting on the fact that numbers do not need to do
this. As you saw from Exercise 2.3, the calculator has a few more
squiggles up its sleeve which might be used for extra numbers.
For example, the numbers could be extended to look like this:

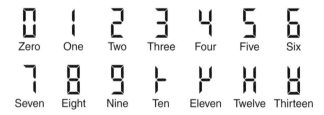

But, as it happens, our number system doesn't look like this.
Probably, for the reason that most humans are born with ten
fingers and ten toes, we have come to an agreement that only

ten unique characters are needed for counting. These are 0, 1, 2, 3, 4, 5, 6, 7, 8 and 9.

These ten characters are called the *numerals*. The word numeral refers to how we write numbers, rather than being concerned with how many things the numbers represent.

After counting as far as 9, we simply group numbers in tens, and count how many tens and how many left over. For example, here is a scattering of just some of the one pound coins that I happened to find when tidying up my daughter's money box.

In order to count them, I might group them in tens, as follows.

So the columns represent the tens, and the ones left over are the units. There are therefore 23 coins here: two tens and three units.

If you feel that you need practice at dealing with tens and units for slightly larger numbers, have a go now at Exercise 2.4.

Exercise 2.4 Practising with tens and units

Switch your calculator on and once again set up the constant 'add 1' by pressing the sequence: ⊞ 1 ⊟ (or by another method given in the subsection 'The calculator constant').

Now enter the number 37 and press the ⊟ key four times.

The result should be 41.

In other words,

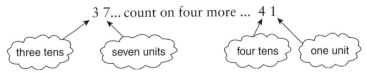

Note that pressing any key at this point, other than the number keys and $\boxed{=}$, is likely to destroy the constant setting. If, for example, you have already pressed the 'clear' key, probably marked $\boxed{\text{c}}$ or $\boxed{\text{ON/C}}$, you will need to reset the constant by rekeying $\boxed{+}$ 1 $\boxed{=}$ or an equivalent key sequence.

Once again, at the risk of being boring, you are reminded not to clear the screen after each sequence, as this will destroy the constant setting.

Enter	Press	Expected tens	Expected units	Answer
44	$\boxed{=}\boxed{=}\boxed{=}$	4	7	47 ←
59	$\boxed{=}\boxed{=}\boxed{=}\boxed{=}\boxed{=}$	6	4	64 ←
73	$\boxed{=}\boxed{=}\boxed{=}\boxed{=}\boxed{=}\boxed{=}\boxed{=}\boxed{=}$	8	1	81 ←
66	$\boxed{=}\boxed{=}\boxed{=}\boxed{=}$	7	0	70 ←

Don't press 'Clear' here

This idea of grouping in tens is the very basis of counting up as far as 99. The next section goes beyond tens and units, into the world of hundreds and thousands.

Hundreds, thousands and beyond

Exercise 2.5 Handling hundreds

Switch your calculator on and once again set up the constant 'add 1' (by a method suitable for your particular machine).

Now enter the number 96 and press the $\boxed{=}$ key four times.

The result should be 100.

In other words,

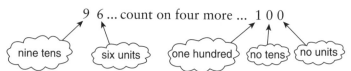

9 6 ... count on four more ... 1 0 0

nine tens   six units   one hundred   no tens   no units

You should not press any key other than the number keys and $=$. As before, if you have destroyed the constant, you will need to reset it by rekeying $+$ 1 $=$ or an equivalent key sequence.

Remember also not to clear the screen after each sequence.

Enter	Press	Expected hundreds	Expected tens	Expected units	Answer
198	$=$ $=$ $=$	2	0	1	201
399	$=$ $=$ $=$ $=$ $=$	4	0	4	404
696	$=$ $=$ $=$ $=$ $=$ $=$	7	0	2	702
897	$=$ sixteen times	9	1	3	913

The largest number you can produce with three figures is nine hundred and ninety-nine. If you want to get any larger, you need to regroup and create a new category of 'ten hundreds', which we call one thousand.

Let's explore thousands by setting the calculator to count up in much larger steps; this time in intervals one hundred at a time. You may be able to work out for yourself how to do this. If not, try one of the following sequences.

Press $+$ 100 and then the $=$ repeatedly.

Or: Press 100 $+$ $+$ and then the $=$ repeatedly.

Or: Press 100 $+$ 100 and then the $=$ repeatedly.

As you do this exercise, try to develop a sense of what to expect at the next press of $=$, particularly when the number in the hundreds column is a 9.

To end this section, it is useful to be able to say numbers as well as to write and interpret them.

Exercise 2.6 Saying numbers out loud

Set your calculator constant to add 1423.

This is done by pressing ⊞ 1423 (or 1423 ⊞ ⊞ or 1423 ⊞ 1423).

Now, each time you press the ⊟ key, try to say out loud the number you see on the screen. Incidentally, it helps if you say the numbers VERY LOUDLY INDEED. Don't worry if other members of your household or your dog think you are crossing the final frontier. This is normal behaviour for mathematical geniuses.

When you have completed this exercise, you can check your answers against those given at the end of the chapter.

The number system represents a central theme in the work of teachers of children in the early years of schooling. You may have children of your own or may be interested in how these ideas are tackled with young children. The final section of this chapter invites you into the infant classroom, to see what things go on, and what sorts of notions and difficulties young children have with numbers.

Exercise 2.7 Ordering your figures

Using the figures 3, 9, 4 and 6 once only, write down:

a the largest possible four-figure whole number

b the smallest possible four-figure whole number.

Beyond thousands, tens of thousands and hundreds of thousands lie two additional words that you should know: millions and billions.

A million is a thousand thousand or 1 000 000.

A billion is a thousand million or 1 000 000 000.

Incidentally, notice how, with very large numbers like these, the digits are often grouped in threes simply to make them easier to read.

Spotlight

If you think a billion is a large number, try a googol, which is 1 followed by one hundred zeros. Put another way, it is equivalent to ten billion, billion, billion, billion, billion, billion, billion, billion, billion, billion, billion.

By the way, the Google website was named as a play on the word 'googol'.

To end this section, the term *place value* needs to be explained. Place value is really another way of describing the whole idea of hundreds, tens and units. It refers to the key idea of how our number system works, namely that the 'place' of a digit (in other words, its position in the number) is what determines its value. For example, the two-digit number 37 is written as 3, followed on the right by 7. By convention, we agree that the first of these digits, the three, refers to three tens while the seven refers to seven units. To understand this point is to understand the principle of 'place value'.

Children and numbers

For most children, number words first come into their world through songs ('Five little speckled frogs sat on a speckled log', 'One, two, three, four, five, once I caught a fish alive', and so on). A feature of many such songs is that the numbers are sung in sequence – sometimes the numbers go up and sometimes they go down. This property of the way that numbers follow on from each other in sequence 1, 2, 3, 4, 5 ... is the *ordinal* property of numbers (ordinal as in the 'order').

The *number line* shown below is a way of helping children to form a mental picture of the sequence of numbers.

However, it is quite a step for children to understand numbers in a number line, to being aware of what the 'three-ness' of three really means in terms of the number of objects – 3 toys, 3 bricks, 3 sweets, 3 calculators, and so on. It is a major breakthrough for a child when she learns that 'three' is a description that can be applied to a whole variety of different collections (or sets) of things. She will need many occasions to grasp three objects physically before three-ness becomes a concept that she can grasp mentally. This idea of a number describing *how many things there are* is known as the *cardinal* property of numbers. Many school activities in the early years deal with this concept by means of play involving counting out various objects – bricks, bottle tops, match boxes, and so on.

Finally, just in case you thought that helping children to understand numbers was a straightforward exercise, try to pick your way through the following exchange between a four-year-old and his teacher, recorded in the book *Wally's Stories* by Vivian Paley.

'We have three 12s in this room,' Wally said one day. 'A round 12, a long 12, and a short 12. The round 12 is the boss of the clock, the long 12 is the ruler, and the short 12 is on a calendar.'

'Why is the 12 on the calendar a short 12?' I asked.

'Me and Eddie measured it. It's really a five. It comes out five on the ruler.'

'Right. It's five.' Wally stared thoughtfully at the clock.

'I'm like the boss of March because my birthday is March 12. The 12 is on the top of the clock.'

Answers to exercises for Chapter 2

Exercises 2.1 to 2.5: No additional comments.

EXERCISE 2.6

Written on the Screen	Said Very Loudly Indeed
1423	one thousand four hundred and twenty-three
2846	two thousand eight hundred and forty-six
4269	four thousand two hundred and sixty-nine
5692	five thousand six hundred and ninety-two

Continued

Written on the Screen	Said Very Loudly Indeed
7115	seven thousand one hundred and fifteen
8538	eight thousand five hundred and thirty-eight
9961	nine thousand nine hundred and sixty-one
11384	eleven thousand three hundred and eighty-four
12807	twelve thousand eight hundred and seven

EXERCISE 2.7

Using the figures 3, 9, 4 and 6:

a the largest possible four-figure whole number is 9643.

b the smallest possible four-figure whole number is 3469.

Summary

In this chapter you were introduced to your calculator and shown its constant facility. You were then shown how to use the *calculator constant* to count in ones or in any interval, and you were then asked to count your way through our whole number system, based on *tens*, *hundreds* and *thousands*. By closely examining how numbers are represented on the calculator display, we looked at the *numerals*, i.e. how numbers are written. But numbers have other features worth exploring. Firstly, they form a natural sequence. This is known as their *ordinal* property and is nicely represented on a *number line*. They are also important as a way of describing 'how many', which is their *cardinal* property.

CHECKLIST

You should have the ability to:

▶ use the constant facility on your calculator to count on or back in any interval.

To find out more via a series of videos, please download our free app, *Teach Yourself Library*, from the App Store or Google Play.

3

Calculating with numbers

In this chapter you will learn:

▶ *some important number words – like prime, square, odd and even*

▶ *what calculations you need in different situations*

▶ *about the 'four rules': add, subtract, multiply and divide*

▶ *how negative (i.e. minus) numbers fit in.*

Properties of numbers

Before tackling the sorts of calculation that we normally do with numbers, it is worth looking at some of the more useful properties of numbers. You will find out what is meant by *odd*, *even*, *prime*, *rectangular* and *square* numbers. These terms are best understood by seeing the numbers arranged in patterns on the table in front of you. So, before reading on, try to get hold of 10 or 12 small identical objects (coins, buttons, or paper clips, for example).

EVEN OR ODD

Take several of your objects (any number between 1 and 12 will do) and try to arrange them into two rows, like this:

If, like me, you chose a number of objects that produced two equal rows, then yours was an even number of objects. In this case, for example, choosing ten objects produced two even rows of five each. So ten is an even number.

However, perhaps your selection didn't work out like this, and when the two equal rows were formed, there was an odd one over – like this:

the odd one over

When any selection of things that are laid out into two rows produce an odd one over, then there must have been an odd number of them. In the last case, for example, choosing 11 objects produced two equal rows of five each, plus an odd one over. So 11 is an odd number.

Exercise 3.1 will give you practice at deciding whether a number is even or odd.

Exercise 3.1 Even or odd?

For each number, mark the box corresponding to whether it is even or odd. The first one has been done for you.

Number	11	7	2	12	8	6	3	1	100
Even	☐	☐	☐	☐	☐	☐	☐	☐	☐
Odd	☒	☐	☐	☐	☐	☐	☐	☐	☐

PRIME, RECTANGULAR AND SQUARE

Now choose six objects. Notice that they can be arranged in a rectangle as two rows of three, like this:

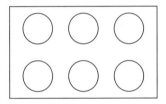

or as three rows of two, like this:

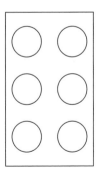

Either way, six is a rectangular number because it *can* be arranged in the form of a rectangle. Similarly, the number 18 is rectangular because 18 objects can be put in this way:

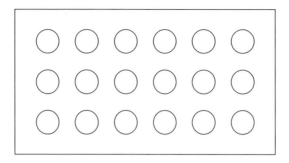

Now try the same task with seven objects. You will soon find that this is impossible. No matter how you move them around, the objects will not form a rectangle; they can only be placed on a line, like this.

So seven is not a rectangular number.

Any number which can't be arranged in the shape of a rectangle is called *prime*, so 7 is a prime number.

Pause for a few moments now and think about what other prime numbers there are.

Now take nine objects and arrange them into three rows and three columns, as shown below.

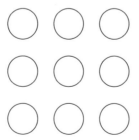

Notice that they have formed a square shape. The number 9 is a square number because it *can* be arranged in the form of a square. (*Note*: 9 is also a rectangular number. A square is a particular type of rectangle.) What other square numbers can you think of?

The next exercise will give you the opportunity to practise your understanding of these three terms: prime, rectangular and square numbers.

Exercise 3.2 Prime, rectangular or square?

For each number, mark the boxes corresponding to whether it is prime, rectangular or square. The first one has been done for you. Note that 9 is both rectangular and square, so two boxes have been marked.

Number	9	7	2	12	8	5	3	4	11
Prime	☐	☐	☐	☐	☐	☐	☐	☐	☐
Rectangular	☒	☐	☐	☐	☐	☐	☐	☐	☐
Square	☒	☐	☐	☐	☐	☐	☐	☐	☐

You have already been introduced to the idea of square numbers like 9 and 25. You should also know that the opposite of a square is called a *square root*, and it is written as $\sqrt{}$. A few examples should illustrate what square root means.

The square of 5, written as 5^2, is 25.
The square root of 25, written as $\sqrt{25}$, is 5.

The square of 9, written as 9^2, is 81.
The square root of 81, written as $\sqrt{81}$, is 9.

The square of 10, written as 10^2, is 100.
The square root of 100, written as $\sqrt{100}$, is 10.

Adding

The four most basic things you can do to any two numbers is add (+), subtract (−), multiply (×) and divide (÷). These are known as the *four rules* (sometimes referred to as the four 'operations').

In the following four sections we will now explain what each operation means and how to perform it. You will be shown how to:

▶ do each operation on the calculator

▶ relate each operation to a helpful mental picture

▶ do the operations mentally (i.e. in your head)

▶ do the operations using pencil and paper.

CALCULATOR ADDITION

Switch your calculator on and, if necessary, press C to clear the screen. Start with some simple 'guess and press' additions as shown in Exercise 3.3.

Exercise 3.3 'Guess and press' additions

a Guess the results to each addition below and then check your answers on the calculator.

3 [+] 7 [=]

7 [+] 3 [=]

12 [+] 9 [=]

99 [+] 5 [=]

b Look again at your answers to the first two questions. Write down in your own words what important property about addition this demonstrates.

c Does the property you noted in part (b) also hold true for subtraction, multiplication and division?

PICTURING ADDITION

Young children's first exposure to addition is normally based on the idea of being given a certain number of a particular object (counters, sweets, coins, ...) and then some more. The question to ponder is: 'How many do I now have altogether?' Children find it helpful to establish these ideas concretely using physical objects.

3 coins plus 2 more makes 5 altogether

Later, they will be shown numbers presented pictorially and the usual image is a number line. Here, addition is represented by a movement to the right.

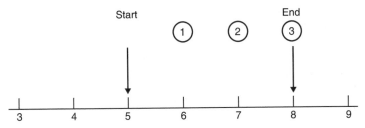

This example shows why 5 + 3 = 8. Start at position 5, take 3 steps to the right and you end at position 8.

MENTAL ADDITION

There is no getting away from it – being able to add is essential, both as a useful everyday skill and as a key building block for other mathematics. As you will see in Exercise 3.4, the calculator can really help you to build up those mental calculating muscles.

Exercise 3.4 Practising adding

Set up your calculator constant to 'add 5'.

Press 0 and then [=] to see a 5 displayed.

Keep pressing [=] repeatedly to see 10, 15, 20, 25 ...

Now, don't press any of the operation keys, or the Clear key, as this will switch off the constant.

a Set up your calculator constant to 'add 4'. Then press 0 [=] [=] [=] ..., but this time try to guess each answer before it is displayed.

b Set up your calculator constant to 'add 7' and repeat the exercise given in part (a). Then, without switching off the constant, key in a different number (say, 2) and repeat by pressing [=] [=] [=] ...

c Now try something more challenging, such as setting the calculator constant to 'add 17' or 'add 23'.

How would you react to adding, say, 19 and 49 in your head? Many people find this hard, but here's a tip.

▶ Add 1 onto each number, giving 20 and 50 (note that you have added 2 extra)

▶ Add these numbers: 20 + 50 = 70

▶ Adjust your answer (i.e. take off the 2 extra): 70 − 2 = 68

You can practise this shortcut by doing Exercise 3.5.

Exercise 3.5 Speed addition

Use the tip shown above to do these additions:

a 28 + 39   **b** 21 + 59   **c** 19 + 38 + 9

Spotlight

Carl Friedrich Gauss (1777–1855) was a precocious mathematical talent from an early age. The story goes that at the age of seven, he and his classmates were asked to add together all the whole numbers from 1 to 100 (presumably an exercise designed by the teacher to keep them quiet for an hour or so). Gauss came up with the answer in a few minutes. He had spotted that by taking the numbers in pairs, 1 and 100, 2 and 99, 3 and 98, and so on, these pairings each added to 101. There were 50 such pairings so the answer must be $50 \times 101 = 5050$.

PENCIL AND PAPER ADDITION

Now, put your calculator to one side, and look at how addition with pencil and paper has traditionally been taught. These written methods are more complicated to explain than to do and, with a bit of practice, you'll quickly speed up. Note that with the pencil and paper method shown here, you always work from right to left.

Example 1

Calculate 27 + 68.

Set out the sum like this with units under units (U) and tens under tens (T).

```
 T U
 2 7
+6 8
```

Add the units (7 + 8 = 15). Note that 15 is 5 units and 1 ten. Write the 5 in the units column and 'carry' the 1 ten over into the tens column (shown here by a small '1' beside the 6).

```
 2 7
+6,8
   5
```

Add the tens (2 + 6 + 1 = 9) and write the 9 in the tens column.

```
 2 7
+6,8
```

The answer: 95 ———————➤
```
 9 5
```

Example 2

Calculate 173 + 269.

Set out the sum like this with units (U), tens (T) and hundreds (H).

```
 H T U
 1 7 3
+2 6 9
```

Add the units (3 + 9 = 12).
Write the 2 in the units column and 'carry' the 1 ten over into the tens column (shown here by a small '1' beside the 6).

```
 1 7 3
+2 6,9
     2
```

Add the tens (7 + 6 + 1 = 14).
Note that 14 tens is really 4 tens and 1 hundred. Write the 4 in the tens column and 'carry' the 1 hundred over into the hundreds column (shown by a small '1' beside the 2).

```
 1 7 3
+2,6,9
   4 2
```

Add the hundreds (1 + 2 + 1 = 4) and write the 4 in the hundreds column.

```
 1 7 3
+2,6,9
```

The answer: 442 ———————➤
```
 4 4 2
```

If you need more practice at pencil and paper addition, try making up some questions, do them and check your answers on the calculator.

Subtracting

CALCULATOR SUBTRACTION

Switch your calculator on and, if necessary, press C to clear the screen. Try the 'guess and press' subtractions in Exercise 3.6.

Exercise 3.6 'Guess and press' subtraction

a Guess the results to each subtraction below and then check your answer on the calculator.

12 [−] 7 [=]
7 [−] 12 [=]
32 [−] 27 [=]
71 [−] 57 [=]

b Look again at your answers to the first two questions. Write down in your own words what important property about subtraction this demonstrates.

PICTURING SUBTRACTION

Children's concrete version of subtraction is based on starting with a particular number of objects and taking some away. The question is: 'How many do I now have left?'

This picture demonstrates a key feature of subtraction – that it is the reverse of addition.

5 coins take away 2 leaves 3

Whereas addition on a number line is represented by a movement to the right, subtraction is represented by a movement to the left.

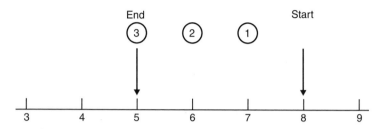

This example shows why $8 - 3 = 5$. Start at position 8, take 3 steps to the left and you end at position 5.

MENTAL SUBTRACTION

Just as you were able to use the calculator constant to help improve addition skills, the same approach can be taken for subtraction.

Exercise 3.7 Practising subtraction

First, set up your calculator constant to 'subtract 5' (use the $\boxed{-}$ rather than the $\boxed{+}$ key). On most basic calculators, you can do this by pressing: 0 [–] 5 [=], or perhaps 5 [–] [–]. Then press 100, and then $\boxed{=}$ to see a 95 displayed. Keep pressing $\boxed{=}$ repeatedly to see the number come down in steps of 5: 90, 85, 80, 75, ...

Now, don't press any of the operation keys or the Clear key as this will switch off the constant.

a Press 84 followed by $\boxed{=}$ $\boxed{=}$ $\boxed{=}$ ... But this time, try to guess each answer before it is displayed.

b Set up your calculator constant to 'subtract 7' and repeat the exercise given in part (a). Then, without switching off the constant, key in a different starting number (say, 215) and repeat by pressing $\boxed{=}$ $\boxed{=}$ $\boxed{=}$ ...

c Now try subtracting larger numbers using the calculator constant.

Note that if you keep subtracting repeatedly, eventually you will reach zero. In fact, it is possible to go below zero, where you will meet up with the negative numbers. Negative numbers are explained at the end of this chapter.

Although pencil and paper calculations are usually done from right to left, there is no rule that requires you to do it this way. For example, suppose you are subtracting 26 from 59 in your head. Here is a useful tip for mental subtraction.

Subtract the tens first: 59 − 20 = 39.

Now subtract the 6: 39 − 6 = 33

Here is a slightly more awkward one: 56 − 29
Subtract the tens first: 56 − 20 = 36
Now subtract the 9: 36 − 9 = 27

You can practise this shortcut now by doing Exercise 3.8.

Exercise 3.8 Speed subtraction

Use the tip shown above to do these subtractions:

a 68 − 35

b 98 − 56

c 72 − 55

Set yourself some more subtractions and check your answers with the calculator.

PENCIL AND PAPER SUBTRACTION

Most children and many adults struggle with performing subtraction on paper. The problem is that it is too easy to simply follow a set of rules. However with no real understanding of why the method works, then you only have to get the rule slightly wrong and the game is up!

There is more than one way of setting out subtraction. However, the method of decomposition is the most common and the easiest to understand. Before galloping into an explanation of decomposition, consider the following scenario. This will explain in everyday terms what decomposition is about.

Suppose you have just bought two sweet packets, each containing 10 sweets – i.e. 20 sweets in all. Six friends turn up and you generously decide to offer a sweet to each one.

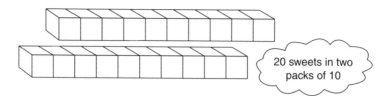

20 sweets in two packs of 10

This involves breaking up one of the packets into 10 single sweets before you can give one to each friend. What you have done here is to decompose one ten into ten separate singles.

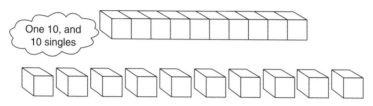

One 10, and 10 singles

After handing over the 6 sweets, you have 14 sweets remaining – one unopened packet of 10 and 4 singles.

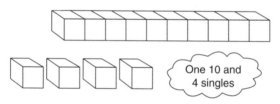

One 10 and 4 singles

Joining in the feeding frenzy, five more friends turn up asking for their sweet. Clearly you can do this but only if you open (i.e. decompose) the second packet. This gives 14 singles.

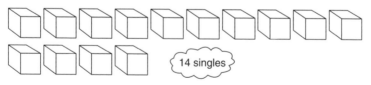

14 singles

You hand over the 5 sweets and are left with 9.

9 remaining singles

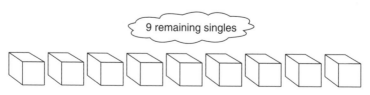

Now look at a written subtraction using this method of 'decomposition'. Note that, as with the written method for addition, you always work from right to left.

Example 3

Calculate 63 − 37.

Set out the sum like this with units under units (U) and tens under tens (T). Try to subtract the units (3 − 7). However, because 7 is bigger than 3, you need to decompose one of the tens in the 63.

```
 T U
 6 3
-3 7
```

Reduce the 6 in the tens column to 5 and 'carry' the 1 ten over into the units column (shown here by a small '1' beside the 3). This gives 13 in the units column.

```
 5¹3
-3 7
```

Subtract the units (13 − 7 = 6) and write the 6 in the units column.

```
 5¹3
-3 7
   6
```

Subtract the tens (5 − 3 = 2) and write the 2 in the tens column.

```
 5¹3
 3 7
```

The answer: 26 ⟶ 2 6

Example 4

Calculate 534 − 147.

Set out the sum in columns like this, marked units (U), tens (T) and hundreds (H). Try to subtract the units (4 − 7). However, because 7 is bigger than 4, you need to decompose one of the tens in the 534.

```
 H T U
 5 3 4
-1 4 7
```

Reduce the 3 in the tens column to 2 and 'carry' the 1 ten over into the units column (shown here by a small '1' beside the 4). This gives 14 in the units column.

```
 5 2¹4
-1 4 7
```

Subtract the units (14 − 7 = 7) and write the 7 in the units column. Try to subtract the tens (2 − 4). However, because 4 is bigger than 2, you need to decompose one of the hundreds in the 534.

```
 5 2¹4
-1 4 7
     7
```

Reduce the 5 in the hundreds column to 4 and 'carry' the 1 hundred over into the tens column (shown here by a '1' beside the 2). This gives 12 in the tens column.	4¹2¹4 1 4 7 7
Subtract the tens (12 − 4 = 8) and write the 8 in the tens column.	4¹2¹4 1 4 7 8 7
Finally, subtract the hundreds (4 − 1 = 3) and write 3 in the hundreds column. The answer: 387 ⟶	4¹2¹4 1 4 7 3 8 7

If you need more practice at pencil and paper subtraction, try making up some questions, do them and check your answers on the calculator.

Multiplying

CALCULATOR MULTIPLICATION

The calculation 3 × 4 can be thought of as three lots of 4. In other words, it is 4, then another 4 and then another 4. So, '3 times' can be thought of as repeatedly adding 4, three times. This is an important link between addition and multiplication – that multiplication can be thought of as repeated addition.

This idea is neatly demonstrated on the calculator using the adding constant. For example, set up the calculator constant to 'add 4' (this was introduced in the subsection on 'The calculator constant' in Chapter 2). Then enter 0, and then press [=] three times. The display goes through the '4 times' table: 4, 8, 12.

Many people have trouble remembering their multiplication tables (9 × 6 and 8 × 7 are particularly troublesome). There are suggestions for improving your multiplication skills in Exercise 3.9.

Exercise 3.9 Developing multiplication skills with a calculator

a Suppose you want to practise your 7 times tables. First, set the calculator constant to 'add 7'. Enter 0 and then repeatedly press ⎡=⎤. Try to guess each value before it is displayed on the screen.

b Without pressing ⎡c⎤ or any other operation key, enter 0 once more and again press ⎡=⎤ repeatedly. This time you should be better at guessing each value before it appears. Do this exercise several times until the numbers that form the '7 times table' are well established in your mind.

c Set the calculator constant to 'times 7'. Enter, say, 5 and guess the result of 7×5. Press ⎡=⎤ to see if you were correct. Now, without pressing ⎡c⎤ or any other operation key, enter, say, 8 and guess the result of 7×8. Again press ⎡=⎤ to see if you were correct. Repeat this exercise until you are getting the correct answer every time.

PICTURING MULTIPLICATION

Building on the idea of repeated addition, young children are offered a vision of multiplication based on rows of identical objects.

For example, 12 can be thought of either as three rows of four, or four rows of three.

Pictures like these help to explain why $3 \times 4 = 4 \times 3$.

MENTAL MULTIPLICATION

There are several useful tricks and tips to improve your multiplication skills. Try these.

a Multiplying by 10: Multiplying a whole number by 10 is easy – just add zero. So $6 \times 10 = 60$, $9 \times 10 = 90$, $115 \times 10 = 1150$, and so on.

b As a useful check on your answer, any multiplication that includes an even number always gives an answer that is an even number. Thus 12×7 or 9×16 will both give an even answer because 12 and 16 are even. Note that even numbers are easy to spot, since their last digit is always even. So, 12, 38, and 126 are even, but 37 and 83 are not.

c Break it in two: Suppose you are multiplying by 6; this is the same as multiplying by 3 and then by 2. For example, to find 7×6, first find 7×3 (which is 21) and then double your answer (42).

d Picture it: Building on the image of multiplication as a rectangle made up of a series of identical rows of objects, it is possible to create two or more rectangles whose areas can be summed to give the answer you require. For example, you may already know that $30 \times 20 = 600$. This can be represented by the first rectangle below, which has an area of 600 (it may help to think of 20 rows each containing 30 coins, giving 600 coins altogether). Now try a harder one: 30×24. You can extend the rectangle by 4 units to form two rectangles. The dimensions of the new one are 30×4, with an area of 120. So the answer to 30×24 is $600 + 120 = 720$.

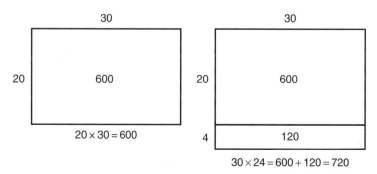

$20 \times 30 = 600$

$30 \times 24 = 600 + 120 = 720$

PENCIL AND PAPER MULTIPLICATION

As was the case for pencil and paper addition and subtraction, the usual written procedure for multiplication is based on working from right to left, starting with the units, then the tens, and then the hundreds.

Example 5

Calculate 54×37.

Set out the calculation like this with units under units (U) and tens under tens (T). I've added the hundreds (H) and thousands (Th) columns as well, as you will need them in this calculation.

```
Th H T U
     5 4
   × 3 7
```

Multiplying by 7 units

Multiply the unit digit (4) by 7 ($4 \times 7 = 28$). Write down the 8 and carry the 2 tens over to the tens column (shown here by a small '2' below the 3).

```
    5 4
  × 3 7
    ₂8
```

Multiply the tens digit (5) by 7 ($5 \times 7 = 35$). Write down the 5 in the tens column and carry the 3 hundreds over to the hundreds column (shown here by a '3' in the previously empty hundred column). This completes the 'multiplying by 7' part.

```
    5 4
  × 3 7
  3 5₂8
```

Multiplying by 3 tens

Note that the 3 is in the tens column so it is actually 3 tens, or 30. The key strategy for multiplying by 30 is to break it into parts. First, multiply by 10 by putting a zero in the units column.

```
    5 4
  × 3 7
  3 5₂8
      0
```

Multiply the unit digit (4) by 3 ($4 \times 3 = 12$). Write down the 2 in the tens column and carry the 1 hundred over to the hundreds column (shown here by a small '1' below the 3).

```
    5 4
  × 3 7
  3 5₂8
   ₁2 0
```

Multiply the tens digit (5) by 3 ($5 \times 3 = 15$). Write down the 5 in the hundreds column and carry the 1 thousand over to the thousands column (shown here by a '1' in the previously empty thousands column).
This completes the 'multiplying by 3 tens' part.

```
     5 4
   × 3 7
   3 5₂8
 1 5₁2 0
```

$$\begin{array}{r} 5\,4 \\ \times\,3\,7 \\ \hline 3\,5_{\,2}8 \\ 1\,5_{\,1}2\,0 \\ \hline 1\,9\,9\,8 \end{array}$$

Finally, add all the sub-totals together (including the small digits).

The answer: 1998 ⟶ 1 9 9 8

If you need more practice at pencil and paper multiplication, try making up some questions, do them and check your answers on the calculator.

Spotlight: Doubling is troubling

Imagine tearing a large sheet of paper in half. Place one half on top of the other. Then tear the two pieces in half again and stack them together, making a pile four pieces high. Repeat the process to produce eight sheets. Continue in this way until you have made 50 tears. Estimate the height of the paper pile. Most people will come up with answers like 10 cm or maybe one metre high. Very few would put the answer at anything like, say, the height of a house. However, even that greatly underestimates the size of the answer. If you try this on a calculator, you will be amazed to learn that, if this exercise were possible (which of course it isn't) the resulting pile of paper would correspond to a journey of roughly 130 trips to the Moon and back!

Dividing

So, to the fourth and final operation described here – division. As you will see, just as subtraction is the reverse of addition, so division is the reverse of multiplication.

CALCULATOR DIVISION

The 'divide' key on a calculator is usually written ⌑÷⌑, although sometimes it is shown as ⌑/⌑ (on a computer, for example). You will have noticed that, if you start off with two whole numbers and multiply them, you always end up with a whole number answer. One reason that people find division hard is that this is not true of division. More often than not, you end up with an answer that is not a whole number, and this will be displayed in decimal form. Dealing with non-whole numbers (fractions, decimals and percentages) is covered in some detail in Chapters 4, 5 and 6.

Exercise 3.10 'Guess and press' division

Guess the solution to each of the calculations below and check your answers on a calculator. (*Note*: they have all been chosen to produce whole number answers.)

a $12 \div 4$

b $12 \div 3$

c $24 \div 8$

d $48 \div 12$

e $72 \div 9$

f $2230 \div 10$

g $52\,400 \div 100$

PICTURING DIVISION

The key to understanding division is the idea of sharing equally. For example, a child might be asked to share, say, 12 sweets fairly among 4 people. How can they all get the same number of sweets?

The answer is to share (or divide) 12 by 4: $12 \div 4 = 3$

Building on the idea of sharing, young children are offered a vision of division based on splitting up a collection of objects. For example, 12 can be shared among 4 people … giving 3 each.

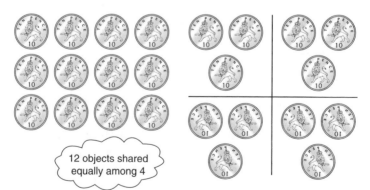

12 objects shared equally among 4

Suppose you are now dividing 14 objects equally among 4 people. The first 12 objects can be allocated (giving 3 each), but what is to be done with the remaining 2? There are two choices – either call these the 'remainder' and leave them unshared, or attempt to split the remainder into fractions. In this example, armed with a strategic hacksaw, you could halve the two remaining coins, so each person gets 'three and a half' ($3\frac{1}{2}$) coins.

MENTAL DIVISION

Here are some tricks and tips to improve your dividing skills.

a   Dividing by 10: If a whole number ends in zero, it can be divided exactly by 10; if it ends in 00, it can be divided by 100. To divide, say, 320 by 10, simply remove the final 0. So $320 \div 10 = 32$. To divide, say, 5700 by 100, remove the final 00. So $5700 \div 100 = 57$.

b   Be aware that all even numbers can be divided by 2 (this is really what the term 'even' means). Thus, 36 is clearly even (since its final digit is even) and so $36 \div 2 = 18$.

c   Break it in two: The tip suggested for multiplication also applies to division. For example, suppose you want to divide 224 by 14. Since 14 is 2×7, you can first divide by 2 ($224 \div 2 = 112$) and then divide by 7 ($112 \div 7 = 16$).

PENCIL AND PAPER DIVISION

The usual written procedures for division are based on working from left to right, starting with the hundreds, then the tens and then the units. This is different from pencil and paper addition, subtraction and multiplication, where you work from right to left.

There are two traditional methods of pencil and paper division, known as 'short' and 'long' division. Long division is usually reserved for calculations involving large numbers. Short and long division calculations are explained below.

Example 6 Short division

Calculate 216 ÷ 9 using short division.
Set out the calculation like this.

$$H\ T\ U$$
$$9)\underline{2\ 1\ 6}$$

Start by dividing the 2 by 9. The answer to 2 ÷ 9 is 0 remainder 2. Write the 0 below the line in the hundreds column and carry the remainder into the next column (shown as a small 2 beside the 1). Since this represents 2 hundreds, there are now 21 tens in the tens column (note that this is another example of decomposition).

$$9)\underline{2\ _2 1\ 6}$$
$$0$$

Divide the 21 tens by 9. The answer to 21 ÷ 9 is 2 remainder 3. Write the 2 below the line in the tens column and carry the remainder into the next column (shown as a small 3 beside the 6). There are now 36 units in the units column.

$$9)\underline{2\ _2 1\ _3 6}$$
$$0\ 2$$

Finally, divide the 36 by 9. The answer to 36 ÷ 9 is 4. Write this below the line in the units column.

The answer: 24 ⟶

$$9)\underline{2\ _2 1\ _3 6}$$
$$0\ 2\ 4$$

If you need practice at pencil and paper division, try making up questions, do them and check your answers on the calculator. If you want to select numbers that divide out exactly, start with two whole numbers (say, 8 and 37) and multiply them on the calculator (giving 296). Then tackle 296 ÷ 8 using short division, knowing that the answer will be a whole number.

As the name suggests, short division is the easier of the two pencil and paper methods of division. It is normally used when the divisor (this is the number you are dividing by) is a single-digit number. But what if you are faced with something like 867÷23? Here the divisor (23) is a two-digit number, and this would make short division a difficult method to use.

Well, personally this is where I'd probably whip out my calculator or turn to my computer to do the calculation. But if you want to do it using pencil and paper, the method of long division would be an appropriate choice. This calculation is shown in Example 7 below.

Example 7 Long division

Calculate 867 ÷ 23.

Set out the calculation like this, with hundreds (H), tens (T) and units (U) marked as shown.

```
  H T U
23)8 6 7
```

Try to divide 23 into the first digit, 8. It goes zero times, so place zero in the hundreds column above the division line.

```
   0
23)8 6 7
```

Try to divide 23 into the first two digits, 86. You need to do some mental arithmetic here. You can work out in your head that 23 × 4 gives 92, which is slightly too big, so the answer is 3. Place 3 in the tens column alongside the 0.

```
   0 3
23)8 6 7
```

Multiply the 23 by the 3 that you have placed in the tens column and write the answer under the 8 and 6. Calculate 3 × 23 is 69, and your workings now look like this.

```
   0 3
23)8 6 7
   6 9
```

Subtract the 69 from 86, giving 17.

```
   0 3
23)8 6 7
   6 9
   1 7
```

Bring down the next digit from the 867, which is the 7, and place it alongside the 17. This gives 177.

```
   0 3
23)8 6 7
   6 9↓
   1 7 7
```

Try to divide 23 into 177. In your head you can work out that 23 × 8 gives 184, which is just too big, so the answer is 7. Place 7 in the units column alongside the 03.

```
   0 3 7
23)8 6 7
   6 9
   1 7 7
```

Multiply the 23 by the 7 that you have just placed in the units column and write the answer under the 177. 7 × 23 is 161, so now the calculation looks like this.

```
   0 3 7
23)8 6 7
   6 9
   1 7 7
   1 6 1
```

Subtract the 161 from 177, giving 16. The answer to this long division calculation is therefore 37 with a remainder of 16, often written as 37 R16.

```
      0 3 7
23 │ 8 6 7
      6 9
    ─────
      1 7 7
      1 6 1
    ─────
        1 6
```

Notice, by the way, that calculations involving short and long division usually give answers that involve a remainder. This is different from a calculator division, which will give a decimal remainder.

Spotlight

There is a so-called rule of arithmetic that says: 'dividing makes smaller'. Do you think this is true always, never or just sometimes? The correct answer is *sometimes*. Try it for some particular cases and you'll start to see when it is true and when it isn't. For example, starting with 8, divide by 2 and the answer, 4, is smaller than 8. Next, starting with 8, try dividing by 1. The answer, 8, is not smaller. The rule is stretched ever further when you try dividing by a fraction or by a negative number. To come up with a correct rule, you need to clarify the circumstances under which it is true – for example:

'Dividing makes smaller' is true when both numbers are positive and the number that you are dividing by is bigger than 1.

Knowing what sum to do

What causes some confusion when using the four rules is that different people use different words to describe them. Most of these terms are listed in Exercise 3.11. See how many you recognize.

Exercise 3.11 The terms used to describe the four rules

Complete the table below, matching each term to the appropriate rule: +, −, × or ÷.

Term	+	−	×	÷
add				
and				
difference				
divide				
from				
goes into				
how many more				
how many less				
less				
minus				
multiply				
plus				
product				
share				
sum				
subtract				
take away				
times				

With a calculator to hand, actually doing a calculation is usually straightforward. The real skills are knowing what sum to do and how to interpret your answer. Here is an illustration.

Example 8

Calculate $5 \times 4 + 7$.

Solution

The calculator sequence is 5 × 4 + 7 = , giving the answer, 27.

Example 9

I have four bottles of milk delivered each weekday and seven at the weekend. How many bottles are delivered in a week?

Solution

The calculation is made up of two parts.

Five *lots of* 4 bottles (one lot for each of the five weekdays) is a multiplication: 5×4

and seven at the weekend is an addition: $+ 7$

So the solution, as before, is 5 $\boxed{\times}$ 4 $\boxed{+}$ 7 $\boxed{=}$, giving the answer, 27 bottles.

Here are some for you to try yourself.

Exercise 3.12 Calculations in context

1 A bus sets out from the depot with 27 people.
   Calculate the number of people on the bus, if:

   a after the first stop, 18 people get on and 9 get off.
   b after the second stop, 12 people get on and 21 get off.
   c after the third stop, 5 people get on and 16 get off.

2 Over eight minutes, the oven temperature rose from 21 degrees Celsius to 205 degrees Celsius.
   a What was the temperature rise?
   b What was the average temperature rise per minute?

3 Denise and Amy are health visitors. Between them they visit 15 patients every day, seven days a week. How many visits will they have made:

   a in a week?
   b during the month of July?
   c in a year?

4 A bottle of wine is said to provide enough for about 7 glasses. If a glass holds about 10 cl, how much does a bottle hold?

Spotlight

Sometimes the wording of the question can fool you into using the wrong calculation. For example, although the word 'share' usually suggests a division calculation, the following question requires a multiplication.

A box of sweets was shared out among 3 children so that they got 18 each. How many sweets were in the box?

An added complication here is the particular choice of the numbers 18 and 3. The fact that 18 happens to be exactly divisible by 3 may make it even more likely that someone who hadn't read the question carefully would do a division.

NEGATIVE NUMBERS

You may have noticed that the subtractions which you have been asked to do so far have been artificially 'set up', so that you have been taking smaller from larger. Usually, subtraction is seen very much in terms of 'taking away' objects. So, if you start off with three objects, you can't take more than three away. However, subtraction doesn't always involve moving objects around. Look at these two examples:

▶ Although I had only £3 in my bank account, I used my credit card to buy an item for £5.

▶ The temperature was 3°C and it dropped a further five degrees overnight.

The banking system hasn't ground to a halt or the thermometer exploded as a result of these events. We simply solve the problem by inventing a new set of numbers less than zero. In bank statements these have the letters O/D (standing for overdrawn) beside them. Usually, however, we just call them *minus*, or *negative* numbers. The number line should really be extended to the left to look like this, so it can show both negative and positive numbers.

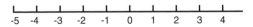

Subtracting 5 from 3 can be shown on the number line as follows:

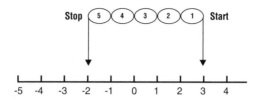

Starting at the number 3 (the right-hand arrow), we take five steps to the left, taking us to the answer -2, said as 'minus two'.

You can check this result by pressing the corresponding key sequence on your calculator. 3 ⬚– 5 ⬚=

Now have a go at the practice exercise below, which should help to consolidate some of the key points of the chapter.

Exercise 3.13 Practice exercise

1 Set yourself a few simple 'sums' using the four rules of +, –, × and ÷. Check your answers using a calculator. Do the 'sums' involving + and – again by drawing a number line and moving, respectively, to the right or left. Check that you get the same answer as with the calculator.

2 The place value of the 6 in the number 365 is <u>ten</u>.
   What is the place value of the digit 6 in the following numbers?

Number	365	614	496	16 042	1 093 461
Place value	ten				

3 7°C is three degrees less than 10°C. Find the temperature which is three degrees less than each of the following.

Temperature °C	10	4	21	-6	-10	0	3	-3
Three degrees less	7							

4 Does it matter in what order you add, subtract, multiply and divide numbers? For example, does 23 ⬚× 15 ⬚= give the same answer as 15 ⬚× 23 ⬚=?

Use your calculator to explore.

5 a Does the square of an even number always give an even number?

   b Does the square of an odd number always give an odd number?

   c Are all odd numbers prime?

   d Are all prime numbers odd?

   e Write down the numbers from 1 to 20, and indicate whether each number is prime, rectangular, odd, even or square.

Answers to exercises for Chapter 3

EXERCISE 3.1

Number	11	7	2	12	8	6	3	1	100
Even			×	×	×	×			×
Odd	×	×					×	×	

EXERCISE 3.2

Number	9	7	2	12	8	5	3	4	11
Prime		×	×			×	×		×
Rectangular	×			×	×			×	
Square	×							×	

EXERCISE 3.3

a 10, 10, 21, 104

b With addition, the order of the numbers being added doesn't matter.

c The property holds true for addition and multiplication, but not for subtraction and division.

EXERCISE 3.4

a 4, 8, 12, 16, 20, … This is the 4 times table.

b 7, 14, 21, 28, 35, … This is the 7 times table.

c 17, 34, 51, 68, 85, … This is the 17 times table.

EXERCISE 3.5

a $28 + 39 = 30 + 40 \, (- 2 - 1) = 70 - 3 = 67$

b $21 + 59 = 20 + 60 \, (+ 1 - 1) = 80 + 0 = 80$

c $19 + 38 + 9 = 20 + 40 + 10 \, (- 1 - 2 - 1) = 70 - 4 = 66$

EXERCISE 3.6

a 5, -5, 5, 14

b You saw in Exercise 3.3b that for addition and multiplication, the order of the numbers being calculated doesn't matter. But with this subtraction exercise, the two answers (5 and -5) were different. This demonstrates that, unlike for addition, the order of the number of matters with subtraction. The order also matters for division.

EXERCISE 3.7

a 84, 79, 74, 69, 64, 59 … Note the repeating pattern in the last digit.

b 84, 77, 70, 63, 56, 49 …
   215, 208, 201, 194, 187 … There is no obvious pattern here.

EXERCISE 3.8

a $68 - 35 = 68 - 30 - 5 = 38 - 5 = 33$

b $98 - 56 = 98 - 50 - 6 = 48 - 6 = 42$

c $72 - 55 = 72 - 50 - 5 = 22 - 5 = 17$

EXERCISE 3.9

No comments

EXERCISE 3.10

a $12 \div 4 = 3$

b $12 \div 3 = 4$ (*Note*: this matches the answer to part a.)

c $24 \div 8 = 3$

d $48 \div 12 = 4$

e $72 \div 9 = 8$

f $2230 \div 10 = 223$

g $52\,400 \div 100 = 524$

EXERCISE 3.11

Term	+	−	×	÷
add	•			
and	•			
difference		•		
divide				•
from		•		
goes into				•
how many more		•		
how many less		•		
less		•		
minus		•		
multiply			•	
plus	•			
product			•	
share				•
sum	•			
subtract		•		
take away		•		
times			•	

EXERCISE 3.12

1 a $27 + 18 - 9 = 36$ people left

   b $\ldots + 12 - 21 = 27$ people left

   c $\ldots + 5 - 16 = 16$ people left

2 a $205 - 21 = 184$ degrees

   b Average temperature rise per minute $= \frac{184}{8} = 23$ degrees

3 a $7 \times 15 = 105$ visits in a week.

   b $31 \times 15 = 465$ visits during the 31 days of July.

   c $365 \times 15 = 5475$ visits in a year.

4 $7 \times 10 = 70$ cl

EXERCISE 3.13

1 No comments

2
Number	365	614	496	16042	1093461
Place value	ten	hundred	unit	thousand	ten

3
Temperature °C	10	4	21	-6	-10	0	3	-3
Three degrees less	7	1	18	-9	-13	-3	0	-6

4 The order doesn't matter when adding and multiplying, but
it does for subtracting and dividing. For example:

Adding $2 + 3 = 3 + 2 = 5$
Multiplying $2 \times 3 = 3 \times 2 = 6$
    BUT
Subtracting $2 - 3 = -1$, whereas $3 - 2 = 1$
Dividing $2 \div 3 = \frac{2}{3}$, whereas $3 \div 2 = 1\frac{1}{2}$

5 a The square of an even number always gives an even
number.

b The square of an odd number always gives an odd number.

c Not all odd numbers are prime (for example, 9, which is
3×3).

d All prime numbers are odd with one exception, namely the
number 2.

e
Number	prime	rectangular	odd	even	square
1			•		•
2	•			•	
3	•		•		
4		•		•	•
5	•		•		
6		•		•	
7	•		•		
8		•		•	
9		•	•		•
10		•		•	
11	•		•		
12		•		•	
13	•		•		

Continued

Number	prime	rectangular	odd	even	square
14		•		•	
15		•	•		
16		•		•	•
17	•		•		
18		•		•	
19	•		•		
20		•		•	

Note: The number 1 has not been marked as either a prime number or a rectangular number. It doesn't comfortably fit into either category, and can't be classified in this way.

Summary

This chapter started by examining some of the properties of numbers – whether they are even or odd, prime, rectangular or square, for example. Next, the so-called 'four rules' of add (+), subtract (−), multiply (×) and divide (÷) were explained on a calculator, in pictures, using mental arithmetic and, finally, using pencil and paper. You were shown how to use a 'number line' to represent numbers and simple calculations. The four rules are, of course, closely connected to each other and these interconnections were explored with the help of a calculator. Finally, you were introduced to negative numbers; these can be thought of as the numbers that appear to the left of zero on the number line.

CHECKLIST
You should have the ability to:

▶ identify even, odd, prime, rectangular and square numbers

▶ add, subtract, multiply and divide using a calculator, mental arithmetic and pencil and paper methods.

To find out more via a series of videos, please download our free app, *Teach Yourself Library*, from the App Store or Google Play.

4

Fractions

In this chapter you will learn:

▶ *how to picture a fraction*
▶ *about equivalent fractions*
▶ *how to calculate with fractions.*

Facing up to fractions

At school, I knew what a fraction was but I just didn't know how to use them. We were told how to add and subtract fractions but I didn't really understand any of it. I never really sorted it out because at my school it was thought to be your fault if you told the teacher that you didn't understand.

Sheila, a friend

Not the most promising start to a chapter on fractions perhaps! It certainly seems to be the case that, while most people know roughly what's going on when the four rules are applied to *whole* numbers, sums with fractions can bring the shutters down. The first thing you should realize is that fractions, decimals and percentages are all very similar. As you will see over the next three chapters, they are all slightly different ways of describing the same thing. What we call fractions – things like $\frac{1}{2}$, $\frac{3}{4}$ and so on – should really be called *common* fractions. In fact these sorts of fractions are not quite as 'common' as they used to be. Increasingly the more awkward common fractions, like $\frac{3}{8}$ and $\frac{5}{32}$, for example, are being replaced by *decimal* fractions. Decimal fractions look like 0.3, 0.125, and so on, and are dealt with in Chapter 5. But first of all, let's find out what a fraction is and where it comes from.

What is a fraction?

Sabine and Sam are four. They have never heard of a fraction. I produced three squares of chocolate and said that they were to be shared between them. They took one square each. Now what about that third square? Well, you can be sure that they won't give it to me, or their favourite charity. Sabine and Sam may not have heard of a fraction, but they are quite capable of inventing one when the occasion arises. As will be explained below, fractions can be thought of as the 'broken bits' that lie between the whole numbers.

Fractions occur quite naturally in division (i.e. sharing) when the sum doesn't divide exactly. For example, sharing 7 doughnuts among 3:

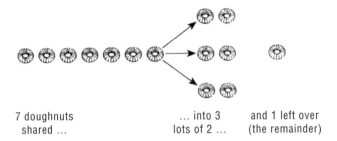

| 7 doughnuts shared ... | ... into 3 lots of 2 ... | and 1 left over (the remainder) |

This can be written as follows:

7/3 = 2 remainder 1

But, as with the square of chocolate, we don't always want to leave the remainder 'unshared'. If the remainder of 1 is *also* shared out among the 3 (people) they each get an extra one third, as shown below.

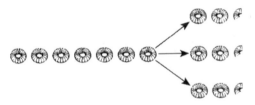

three lots of 'two and one third'

So, the more complete answer to this division sum is:

7/3 = $2\frac{1}{3}$, i.e. 7 divided by 3, gives 2 and a third

It is important to understand why fractions are written as they are. The fraction $\frac{1}{3}$ really is another way of writing 1 divided by 3, or 1/3. So the top number in a fraction (called the *numerator*) is the number of things to be shared out. The bottom number (the *denominator*) tells you how many shares there will be.

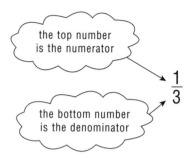

the top number is the numerator

the bottom number is the denominator

$$\frac{1}{3}$$

Exercise 4.1 will help you grasp this important idea.

Exercise 4.1 Sharing cheese

A box of processed cheese has six segments. Share two boxes equally among three people.

a How many segments will each person get?

b What fraction of a box will each person receive?

How to picture a fraction

The common fractions like a half, three quarters and two thirds are part of everyday language. You should find it helpful to have a mental picture of a fraction. The picture which is in my mind (and which is used in most schools when fractions are first introduced) is to imagine a whole as a complete cake. This can be cut into slices representing various fractions, like this:

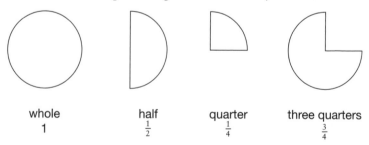

whole	half	quarter	three quarters
1	$\frac{1}{2}$	$\frac{1}{4}$	$\frac{3}{4}$

These sorts of pictures are helpful as a way of understanding what a fraction is. But if you need to compare fractions or do calculations with them, then you need more than pictures. For example:

Is $\frac{2}{3}$ of a cake bigger than $\frac{3}{4}$ of it?

Spotlight

A few years ago, working with primary school teacher Louise Graham, we carried out a piece of 'action research' based on introducing fractions to her pupils. We asked each child to create their own personal fraction kit, consisting of different-sized slices of circles. What was unique about our approach was that we didn't tell the children how big to make each slice – this was something that they had to figure out for themselves. For example, to make a slice of one third, they had to find a way of cutting a cardboard circle up into three identical slices; to make quarters they needed to make four identical slices, and so on.

The outcome of this initiative was remarkable. By the end of the sequence of lessons, every child in the class seemed to have a confident understanding of what a fraction was – something that many pupils never achieve after many conventional lessons on fractions. We concluded that there were two key factors to the success of this work. First, their understanding was based on physicality – they benefited from seeing, touching and picking up actual physical slices of circles. Second, they had to think for themselves, and explain to others, how to make these slices, which we believe gave them a core understanding of fractions that remained with them long afterwards.

This initiative has been written up in several journals, including *Mathematics Teaching*: 'DIY fraction pack', NCETM (National Centre for Excellence in Teaching Mathematics) support CD for teachers, 2008, Alan Graham, Louise Graham.

Fitting fractions into the number line

The next step is to understand how fractions fit into the sequence of numbers that you looked at in Chapter 2. For example:

$2\frac{1}{3}$ is one third of the way between 2 and 3

$3\frac{7}{10}$ is seven tenths of the way between 3 and 4

The diagram below shows how these fractions fit on the number line.

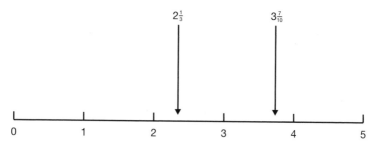

Until fractions are introduced, numbers can be thought of as a set of points equally spaced on the line (i.e. the whole numbers). But as your picture of numbers expands, you can see that there are lots of other points between 1 and 2, between 2 and 3, and so on. How many are there? Are there any gaps at all on the number line when the fractions are added? These aren't questions with easy answers, but you might care to think about them.

Finally, here is a reminder of how the cake diagram and the number line can help your mental picture of fractions.

▶ Cake diagram → fractions are 'bits' of a whole

▶ Number line → fractions fill in the gaps 'between whole numbers'

And now you're ready to add and subtract fractions. Well, nearly... Before that it would be useful to know what *equivalent* fractions are.

What are equivalent fractions?

What is the difference between sharing two cakes among four people or sharing one cake between two people? Well, since everybody ends up with half a cake, there is no difference in the share that each person gets. The first lot of people might actually get $\frac{2}{4}$ of a cake but that would seem to be the same as $\frac{1}{2}$. So $\frac{2}{4}$ and $\frac{1}{2}$ are fractions which are the same; yet they are different – they have the same value but have different numerals top and bottom. The word used to describe this is *equivalence*.

We would say that $\frac{2}{4}$ and $\frac{1}{2}$ are *equivalent fractions*.

Exercise 4.2 (b) will give you a chance to spot some more equivalent fractions. Try it now.

Exercise 4.2 Finding equivalent fractions

a Mark with an arrow each of these numbers on the number line below: $2\frac{3}{4}, \frac{1}{2}, 4\frac{9}{10}, 1\frac{1}{3}$

b Find three fractions equivalent to each of the following (the first set has been done for you):

Fraction	Equivalent fractions
$\frac{3}{4}$	$\frac{6}{8}, \frac{9}{12}, \frac{12}{16}$
$\frac{1}{2}$	
$\frac{9}{10}$	
$\frac{1}{3}$	
$\frac{6}{24}$	
$\frac{10}{20}$	

Adding and subtracting fractions

When adding fractions, it is helpful to think of the slices of cake. For example:

$\frac{1}{8} + \frac{3}{8} = ?$

Here the slices are all the same size ($\frac{1}{8}$ each) so we just add them together, like this:

$$\frac{1}{8} + \frac{3}{8} = \frac{4}{8}$$

It is usual to write the answer in the form of the simplest equivalent fraction, so the answer $\frac{4}{8}$ can be written as $\frac{1}{2}$.

However, what happens when you have to add fractions like the following?

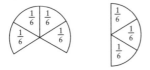

$$\frac{2}{3} + \frac{1}{2} = ?$$

This time the slices of cake aren't the same size, so we can't just add them together. The way out of this problem is to cut both fractions until all the slices are the same size – like this:

Now, with all the slices equal to $\frac{1}{6}$, they *can* be added. The calculation looks like this:

$$\frac{2}{3} + \frac{1}{2}$$
$$= \frac{4}{6} + \frac{3}{6}$$
$$= \frac{7}{6}$$
$$= 1\frac{1}{6}$$

Both fractions are changed to equivalent 'sixths'

But why did I choose to subdivide each fraction into slices of $\frac{1}{6}$? The reason is that $\frac{1}{6}$ is the easiest fraction that a half and a third will break up into. I could have used slices of $\frac{1}{12}$ or $\frac{1}{18}$ but that would have been unnecessarily complicated. By the way, this process of breaking fractions up into smaller slices so that they can be added or subtracted is called 'finding a common denominator'.

Spotlight

It's easy to add and subtract fractions with the same denominators. For example, suppose you are adding $\frac{1}{5}$ and $\frac{2}{5}$; the answer is $\frac{3}{5}$. Here your fractions are both in the same 'units' of fifths. This is like adding 1 metre and 2 metres to get 3 metres – because the measurements share the same units, metres, you simply add the 1 and the 2.

However, suppose you are asked to add 1 metre and 2 feet. You can't do it because the units are different. You would have to convert the units so they are the same, and then you could add them. Exactly the same is true with fractions – you can't add or subtract fractions unless they are composed of the same-sized slice (i.e. unless they share the same denominator).

To summarize, finding the lowest common denominator means finding the smallest number which both the denominators will divide into. (Remember that the denominator is the bottom number in the fraction.) This number then becomes the new denominator. Thus in the example above, the lowest number that 3 and 2 both divide into is 6, so 6 is the new denominator. Now try Exercise 4.3 (the first one has been done for you).

Exercise 4.3 Adding and subtracting fractions

Complete the table below.

Calculation	Equivalent fractions	Answer
$\frac{2}{3} + \frac{1}{4}$	$\frac{8}{12} + \frac{3}{12}$	$\frac{11}{12}$
$\frac{1}{2} + \frac{3}{4}$		
$\frac{1}{3} + \frac{5}{6}$		
$\frac{4}{5} - \frac{1}{2}$		
$\frac{1}{5} + \frac{1}{4}$		
$\frac{1}{4} - \frac{1}{5}$		

Multiplying and dividing fractions

To be honest, there aren't very many practical situations where multiplying and dividing fractions crop up. One instance might be scaling up or down the quantities in a recipe – when you want to produce a smaller or larger cake than the one in the recipe – in this case it is certainly useful to know a few basic facts about performing simple calculations with fractions. For example, you should know that half of a half is a quarter and a tenth of a tenth is one hundredth. Nevertheless, for completeness, this section will cover the basics of multiplying and dividing fractions. One point worth noting here is that, where possible, it is good practice to write fractions in their simplest terms – for example, don't give an answer like $\frac{6}{8}$, but instead reduce this to $\frac{3}{4}$.

MULTIPLYING FRACTIONS

To multiply two simple fractions:

▶ rewrite the fractions as a single fraction

▶ look for opportunities to cancel down the numbers

▶ multiply the two numerators (the top numbers) and the two denominators (the bottom numbers), giving your answer as a fraction.

Example	Calculation
Multiply $\frac{1}{2}$ and $\frac{2}{3}$	$\frac{1}{2} \times \frac{2}{3}$
Rewrite the multiplication as a single fraction.	$\frac{1 \times 2}{2 \times 3}$
If possible, cancel down – in this case you can cancel down by dividing top and bottom by 2.	$\frac{1 \times \cancel{2}1}{\cancel{1}2 \times 3}$
Multiply out the fraction.	$\frac{1 \times 1}{1 \times 3} = \frac{1}{3}$

Example	Calculation
Multiply $\frac{3}{4}$ and $\frac{7}{12}$	$\frac{3}{4} \times \frac{7}{12}$
Rewrite the multiplication as a single fraction.	$\frac{3 \times 7}{4 \times 12}$
If possible, cancel down – in this case you can cancel down by dividing top and bottom by 3.	$\frac{1\cancel{3} \times 7}{4 \times \cancel{12}4}$
Multiply out the fraction.	$\frac{1 \times 7}{4 \times 4} = \frac{7}{16}$

▶ **Example: multiply** $3\frac{3}{4}$ **and** $2\frac{1}{3}$

Solution: This example is more complicated because these two numbers are not simple fractions but instead, *mixed fractions*. A mixed fraction consists of a whole number part and a fraction part. For example, $3\frac{3}{4}$ has a whole number part of 3 and a fraction part of $\frac{3}{4}$. In order to multiply a mixed fraction, it has to be rewritten as a 'vulgar' fraction (sometimes referred to as a 'top-heavy' fraction), which means that the entire number must be expressed as a simple fraction (in the case of $3\frac{3}{4}$, with a denominator of 4). The two numbers are therefore adjusted to vulgar fraction form as follows:

$$3\tfrac{3}{4} = \frac{12}{4} + \frac{3}{4} = \frac{12+3}{4} = \frac{15}{4}$$

$$2\tfrac{1}{3} = \frac{6}{3} + \frac{1}{3} = \frac{6+1}{3} = \frac{7}{3}$$

Now that the numbers have been converted to vulgar fraction form, the calculation can be completed, as follows.

Example	Calculation
Multiply $3\frac{3}{4}$ and $2\frac{1}{3}$	$3\frac{3}{4} \times 2\frac{1}{3}$
Express both numbers as vulgar fractions.	$\frac{15}{4} \times \frac{7}{3}$
Write it as a single fraction.	$\frac{15 \times 7}{4 \times 3}$
Cancel down by dividing top and bottom by 3.	$\frac{5\,\cancel{15} \times 7}{4 \times \cancel{3}\,1}$
Multiply out the fraction.	$\frac{5 \times 7}{4 \times 1} = \frac{35}{4}$
Obey the usual convention by expressing the answer as a mixed fraction.	$8\frac{3}{4}$

DIVIDING FRACTIONS

Before stating the rule for dividing fractions, think about the following situation.

You have three pizzas. How many portions will you have if you divide them into half-pizzas? As the picture shows, the answer is six.

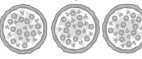   Six half-pizzas

This suggests that 3 divided by $\frac{1}{2}$ gives the answer 6.

Similarly, 3 divided by $\frac{1}{4}$ = 12.

Now, here is the rule for dividing by a fraction.

To divide by a fraction, turn the fraction upside down and then multiply.

To take the pizza example:

Example	Calculation
Divide 3 by $\frac{1}{2}$	$3 \div \frac{1}{2}$
Turn the number that you are dividing by upside down and multiply.	$3 \times \frac{2}{1} = \frac{6}{1} = 6$

Some more examples are given below.

Example	Calculation
Divide $\frac{1}{2}$ by $\frac{2}{3}$	$\frac{1}{2} \div \frac{2}{3}$
Turn the number that you are dividing by upside down and multiply.	$\frac{1}{2} \times \frac{3}{2} = \frac{1 \times 3}{2 \times 2} = \frac{3}{4}$

Example	Calculation
Divide $\frac{3}{4}$ by $\frac{7}{12}$	$\frac{3}{4} \div \frac{7}{12}$
Turn the number that you are dividing by upside down and multiply.	$\frac{3}{4} \times \frac{12}{7} = \frac{3 \times 12}{4 \times 7}$
Cancel down by dividing top and bottom by 4 and multiply out.	$\frac{3 \times \cancel{12}3}{\cancel{14}7 \times 7} = \frac{3 \times 3}{1 \times 7} = \frac{9}{7}$
Give the answer as a mixed fraction.	$\frac{9}{7} = 1\frac{2}{7}$

▶ **Example: divide $1\frac{3}{4}$ by $2\frac{1}{3}$**

Solution: This time, the two numbers are not simple fractions but instead, *mixed fractions*, so as was the case with multiplying mixed fractions, they must first be turned into vulgar fractions as follows:

$$1\tfrac{3}{4} = \tfrac{4}{4} + \tfrac{3}{4} = \tfrac{4+3}{4} = \tfrac{7}{4}$$

$$2\tfrac{1}{3} = \tfrac{6}{3} + \tfrac{1}{3} = \tfrac{6+1}{3} = \tfrac{7}{3}$$

The rest of the calculation is shown in the table below.

Example	Calculation
Divide $1\tfrac{3}{4}$ by $2\tfrac{1}{3}$	$1\tfrac{3}{4} \div 2\tfrac{1}{3}$
Express both numbers as vulgar fractions.	$\tfrac{7}{4} \div \tfrac{7}{3}$
Turn the number that you are dividing by upside down and multiply.	$\tfrac{7}{4} \times \tfrac{3}{7}$
Express it as a single fraction.	$\tfrac{7 \times 3}{4 \times 7}$
Cancel down by dividing top and bottom by 7 and multiply out.	$\tfrac{\cancel{7} \times 3}{4 \times \cancel{7}1} = \tfrac{1 \times 3}{4 \times 1} = \tfrac{3}{4}$

Spotlight

The earliest known use of fractions is around 5000 years ago in the Indus Valley, which is part of Pakistan today. Two thousand years later, the Greeks invented their own form of fractions based on adding *unit* fractions (i.e. fractions with numerators equal to 1). For example, the fraction $\tfrac{3}{4}$ could be expressed as $\tfrac{1}{2} + \tfrac{1}{4}$. Another example is rewriting $\tfrac{11}{12}$ as $\tfrac{1}{2} + \tfrac{1}{3} + \tfrac{1}{12}$.

Come forward another 500 years to the Greek philosophers, the Pythagoreans, who were also very aware of fractions. They discovered that the square root of two (written today as $\sqrt{2}$) could not be expressed as a fraction. This 'unholy' number challenged so many of their beliefs about religion and the natural order of things that they took a vow of secrecy about it, lest such numbers should come to the attention of a wider public. The story goes that one member of the brotherhood let the cat out of the mathematical bag, and was promptly put to death for his indiscretion.

Ratio and proportion

A ratio is a way of describing how something should be shared out into two or more shares. Typically, it is written as two numbers, separated by a colon. So, a simple ratio might be written as something like 2:1 (said as 'in the ratio two to one'). For example, suppose you wish to share £30 between two people in the ratio 2:1. This means that one person gets two shares while the other gets one. The best way of tackling problems like this is to say that there are three shares altogether. Each one is worth £10 (i.e. £30 ÷ 3), so one person gets £20 while the other gets £10.

Here is a slightly harder example. Suppose three children are left £3000 in a will, with the stipulation that it is to be divided in the ratio 1:2:3. To work out what each child gets, you first add the numbers (1 + 2 + 3 = 6). So there are six shares altogether. Divide £3000 by 6 to find that each share is worth £500. So the three children receive £500 (one share), £1000 (two shares) and £1500 (three shares), respectively.

Proportion crops up when a particular ratio is seen to apply to two different situations. For example, in the first triangle below, the longest side is exactly three times as long as the shortest side. Now imagine that you have enlarged this triangle so that its shape stays the same but every length has been doubled. What you find, is that the longest side is still three times as long as the shortest side. This captures the fundamental idea *in proportion* – when the two ratios are the same, the two shapes are in proportion.

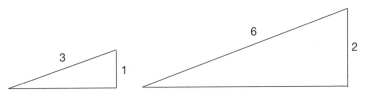

Exercise 4.4: Practice exercise

1   Share the following equally. (The first one has been done for you.)

   a 11 cakes among 4 people. Each gets $2\frac{3}{4}$ cakes

   b 17 cakes among 5 people. Each gets ☐ cakes

   c 5 cakes among 6 people. Each gets ☐ cakes

   d 20 cakes among 3 people. Each gets ☐ cakes

2 The number 1 consists of 3 thirds. How many thirds are there in the following numbers?

    a 2   **d** 12

    b 4   **e** $3\frac{1}{3}$

    c 10   **f** $7\frac{2}{3}$

3 Write the appropriate fractions onto the slices of the clock.

Now add the fractions together.

Check that they add to 1.

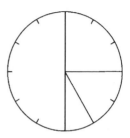

4 The fraction $\frac{8}{10}$ can be written more simply as $\frac{4}{5}$. Write the following fractions in their simplest form. (One has been done for you.)

Fractions	$\frac{8}{10}$	$\frac{4}{6}$	$\frac{5}{10}$	$\frac{12}{18}$	$\frac{6}{9}$	$\frac{4}{16}$	$\frac{8}{48}$	$\frac{9}{18}$	$\frac{2}{22}$
Simplest form	$\frac{4}{5}$								

5 **i** Change all the fractions below to twelfths. (The first one has been done for you.)

    ii Now rank them in order of size, putting a rank of 1 against the largest fraction and 6 against the smallest. (Again, one has been done for you.)

Fractions	$\frac{2}{3}$	$\frac{3}{4}$	$\frac{2}{6}$	$\frac{7}{12}$	$\frac{5}{6}$	$\frac{1}{2}$
Fractions as twelfths	$\frac{8}{12}$					
Rank	3					

6 A sum of £600 000 was left to be shared among three charities, as follows:

Charity A was to receive one quarter.
Charity B was to receive two thirds.
Charity C was to receive the rest.

Calculate:

    a the fraction of the sum that went to Charity C

    b the amount of money due to each charity.

7 Calculate the following:

    a $\frac{1}{2} + \frac{3}{5}$

    b $\frac{1}{2} - \frac{1}{5}$

    c $\frac{3}{4} \times \frac{2}{9}$

    d $\frac{3}{4} \div \frac{2}{9}$

    e $2\frac{1}{4} \times 4\frac{1}{2}$

    f $2\frac{1}{4} \div 4\frac{1}{2}$

Answers to exercises for Chapter 4

EXERCISE 4.1

a Each person gets $\frac{6+6}{3} = \frac{12}{3} =$ 4 segments.

b Expressed as a fraction, each person gets $\frac{4}{6}$ or, in other words, $\frac{2}{3}$ of a box.

EXERCISE 4.2

a

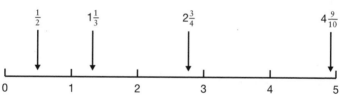

b There are many possible answers here.

Fraction	Equivalent fractions
$\frac{3}{4}$	$\frac{6}{8}, \frac{9}{12}, \frac{12}{16}$
$\frac{1}{2}$	$\frac{2}{4}, \frac{5}{10}, \frac{7}{14}$
$\frac{9}{10}$	$\frac{18}{20}, \frac{45}{50}, \frac{90}{100}$
$\frac{1}{3}$	$\frac{2}{6}, \frac{20}{60}, \frac{30}{90}$
$\frac{6}{24}$	$\frac{1}{4}, \frac{2}{8}, \frac{100}{400}$
$\frac{10}{20}$	$\frac{1}{2}, \frac{9}{18}, \frac{12}{24}$

EXERCISE 4.3

Calculation	Equivalent fractions	Answer
$\frac{2}{3}+\frac{1}{4}$	$\frac{8}{12}+\frac{3}{12}$	$\frac{11}{12}$
$\frac{1}{2}+\frac{3}{4}$	$\frac{2}{4}+\frac{3}{4}$	$\frac{5}{4}=1\frac{1}{4}$
$\frac{1}{3}+\frac{5}{6}$	$\frac{2}{6}+\frac{5}{6}$	$\frac{7}{6}=1\frac{1}{6}$
$\frac{4}{5}-\frac{1}{2}$	$\frac{8}{10}-\frac{5}{10}$	$\frac{3}{10}$
$\frac{1}{5}+\frac{1}{4}$	$\frac{4}{20}+\frac{5}{20}$	$\frac{9}{20}$
$\frac{1}{4}-\frac{1}{5}$	$\frac{5}{20}-\frac{4}{20}$	$\frac{1}{20}$

EXERCISE 4.4

1 a 11 cakes among 4 people. Each gets $\boxed{2\frac{3}{4}}$ cakes

   b 17 cakes among 5 people. Each gets $\boxed{3\frac{2}{5}}$ cakes

   c 5 cakes among 6 people. Each gets $\boxed{\frac{5}{6}}$ cakes

   d 20 cakes among 3 people. Each gets $\boxed{6\frac{2}{3}}$ cakes

2 a $2=\frac{6}{3}$ d $12=\frac{36}{3}$

   b $4=\frac{12}{3}$ e $3\frac{1}{3}=\frac{10}{3}$

   c $10=\frac{30}{3}$ f $7\frac{2}{3}=\frac{23}{3}$

3 The fractions are $\frac{1}{4},\frac{1}{6},\frac{1}{12}$ and $\frac{1}{2}$.

   These can be rewritten in twelfths and added, as follows.

   $\frac{3}{12}+\frac{2}{12}+\frac{1}{12}+\frac{6}{12}=\frac{3+2+1+6}{12}=\frac{12}{12}=1$

4 Fractions

$\frac{8}{10}$	$\frac{4}{6}$	$\frac{5}{10}$	$\frac{12}{18}$	$\frac{6}{9}$	$\frac{4}{16}$	$\frac{8}{48}$	$\frac{9}{18}$	$\frac{2}{22}$

   Simplest form

$\frac{4}{5}$	$\frac{2}{3}$	$\frac{1}{2}$	$\frac{2}{3}$	$\frac{2}{3}$	$\frac{1}{4}$	$\frac{1}{6}$	$\frac{1}{2}$	$\frac{1}{11}$

5 Fractions

$\frac{2}{3}$	$\frac{3}{4}$	$\frac{2}{6}$	$\frac{7}{12}$	$\frac{5}{6}$	$\frac{1}{2}$

   Fractions as twelfths

$\frac{8}{12}$	$\frac{9}{12}$	$\frac{4}{12}$	$\frac{7}{12}$	$\frac{10}{12}$	$\frac{6}{12}$

   Rank

3	2	6	4	1	5

6 **a** Charity C was to receive $1 - (\frac{1}{4} + \frac{2}{3}) = 1 - \frac{11}{12} = \frac{1}{12}$

   b Charity A was to receive $\frac{1}{4} \times £600\,000 = £150\,000$
   Charity B was to receive $\frac{2}{3} \times £600\,000 = £400\,000$
   Charity C was to receive $\frac{1}{12} \times £600\,000 = £50\,000$

   (As a quick check, these amounts of money should add to
   $£600\,000$. $£150\,000 + £400\,000 + £50\,000 = £600\,000$.)

7 **a** $\frac{1}{2} + \frac{3}{5} = \frac{5+6}{10} = \frac{11}{10} = 1\frac{1}{10}$

   b $\frac{1}{2} - \frac{1}{5} = \frac{5-2}{10} = \frac{3}{10}$

   c $\frac{3}{4} \times \frac{2}{9} = \frac{3 \times 2}{4 \times 9} = \frac{1\cancel{3} \times \cancel{2}1}{2\cancel{4} \times \cancel{9}3} = \frac{1 \times 1}{2 \times 3} = \frac{1}{6}$

   d $\frac{3}{4} \div \frac{2}{9} = \frac{3}{4} \times \frac{9}{2} = \frac{3 \times 9}{4 \times 2} = \frac{27}{8} = 3\frac{3}{8}$

   e $2\frac{1}{4} \times 4\frac{1}{2} = \frac{9}{4} \times \frac{9}{2} = \frac{9 \times 9}{4 \times 2} = \frac{81}{8} = 10\frac{1}{8}$

   f $2\frac{1}{4} \div 4\frac{1}{2} = \frac{9}{4} \div \frac{9}{2} = \frac{9}{4} \times \frac{2}{9} = \frac{9 \times 2}{4 \times 9} = \frac{1\cancel{9} \times \cancel{2}1}{2\cancel{4} \times \cancel{9}1} = \frac{1 \times 1}{2 \times 1} = \frac{1}{2}$

Summary

▶ Fractions can be thought of as bits of whole numbers.

▶ A useful way of representing fractions is as slices of a cake.

▶ Equivalent fractions, like $\frac{1}{2}$ and $\frac{2}{4}$ have the same value and correspond to the same size of slice of the cake.

▶ Adding and subtracting fractions usually involves rewriting the fractions as equivalent fractions. This means finding a common denominator (the bottom number in the fraction), and adding the numerators (the top numbers in the new fractions).

▶ Multiplying and dividing fractions is easiest to understand when the fractions are written as decimal fractions (see Chapter 5).

To find out more via a series of videos, please download our free app, *Teach Yourself Library*, from the App Store or Google Play.

5

Decimals

In this chapter you will learn:

▶ *about the 'ten-ness' of numbers*

▶ *why we use a decimal point and where we put it*

▶ *about the connection between fractions and decimals*

▶ *how to calculate with decimals.*

Five fingers on each hand (well, four fingers and one thumb) seems to be a reasonable number to possess. Any fewer and we wouldn't be able to play the 'Moonlight Sonata' with the same panache any more, and we'd have a bit of a struggle putting on a pair of gloves. What I'm really saying, then, is that the reason our number system is based on the number ten is because humans have counted on their ten fingers for thousands of years. 'Decimal' (from the Latin *deci* meaning ten) is really a way of describing the ten-ness of our counting system. However, it usually refers to decimal fractions. And there is no shortage of those around us. Just listen to sports commentators, for example:

> *... the winning time of 9.58 seconds smashes the world record by two hundredths of a second.*
> *... the winning scores for the pairs ice skating are as follows: 5.9, 5.8, 5.9 ...*

Decimal points appear whether we are talking about money (£8.14) or measurement (2.31 metres) and will appear on a calculator display at the touch of a button.

Decimal fractions

A decimal fraction is simply another way of writing a common fraction. In this section, you can use your calculator to discover how fractions and decimals are connected.

Exercise 5.1 Deriving decimals from fractions

For each group of questions below:
a write down the answers in fractions
b use your calculator to find the answers in decimals
c complete the blank in the 'Conclusion' box for each group.

The first one in the group has been started for you.

Key sequence	Fraction	Decimal	Conclusion
1 ÷ 2 =	$\frac{1}{2}$	0.5	
2 ÷ 4 =			
5 ÷ 10 =			
50 ÷ 100 =			The decimal for $\frac{1}{2}$ is [0.5]

$1 \boxed{\div} 4 \boxed{=}$

$2 \boxed{\div} 8 \boxed{=}$

$5 \boxed{\div} 20 \boxed{=}$

$25 \boxed{\div} 100 \boxed{=}$ The decimal for $\frac{1}{4}$ is $\boxed{}$

$3 \boxed{\div} 4 \boxed{=}$

$6 \boxed{\div} 8 \boxed{=}$

$75 \boxed{\div} 100 \boxed{=}$ The decimal for $\frac{3}{4}$ is $\boxed{}$

$1 \boxed{\div} 10 \boxed{=}$

$10 \boxed{\div} 100 \boxed{=}$ The decimal for $\frac{1}{10}$ is $\boxed{}$

As can be discovered from the key sequences above, converting from fractions to decimal fractions is very straightforward using a calculator. For example, the fraction $\frac{5}{8}$ can be converted to a decimal fraction by dividing 5 by 8, i.e. by pressing the following key sequence:

$5 \boxed{\div} 8 \boxed{=}$

This produces the result 0.625.

In other words, the fraction $\frac{5}{8}$ has the same value as the decimal fraction 0.625. Exercise 5.2 will give you practice at converting from fractions to decimal fractions.

Exercise 5.2 Converting from fractions to decimal fractions

Now use your calculator to find the decimal values of the fractions below.

Fraction	$\frac{1}{2}$	$\frac{1}{4}$	$\frac{3}{4}$	$\frac{1}{10}$	$\frac{1}{5}$	$\frac{2}{5}$	$\frac{3}{10}$	$\frac{9}{10}$	$\frac{1}{20}$	$\frac{1}{8}$	$\frac{1}{3}$
Decimal											

Spotlight

PICTURING DECIMAL FRACTIONS

You may remember from the previous chapter, that fractions could be helpfully represented using slices of a cake. Since there is such a close link between fractions and decimal fractions, it follows that the same helpful pictures apply to decimal fractions.

Here are the 'cakes' from Chapter 4, but this time with the corresponding decimal fractions added.

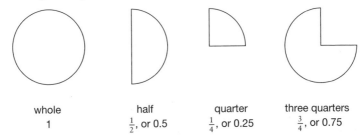

whole	half	quarter	three quarters
1	$\frac{1}{2}$, or 0.5	$\frac{1}{4}$, or 0.25	$\frac{3}{4}$, or 0.75

Let's now turn to the way we represent decimal fractions on a number line. Again, since fractions and decimals are really very similar, it is not surprising that they can both be represented in the same way. For example, the fraction $\frac{3}{4}$ and the decimal 0.75 share the same position on the number line. Thus:

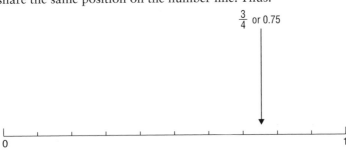

$\frac{3}{4}$ or 0.75

0   1

Having made a connection between fractions and decimals, the next exercise (called 'Guess and press') gets you working just with decimals. The idea is to write down your *guess* as to what the answer will be for each calculation. Then you *press* the key sequence on the calculator and see if you are right. The aim of this exercise is to help you see the connection between decimals and whole numbers.

Exercise 5.3 Guess and press

Calculation	Guess	Press
0.5 [+] 0.5 [=]	1	1
0.5 [×] 2 [=]		
0.25 [×] 4 [=]		
0.5 [×] 10 [=]		
4 [÷] 10 [=]		
0.1 [+] 0.1 [=]		
0.1 [×] 10 [=]		

What is the point of the decimal point?

If the world contained only whole numbers, we would never need a decimal point. However, it is helpful to be aware that decimal numbers are simply an extension of the whole number system. If we think of a whole number, the rules of place value tell us what each of these digits represents. Thus, the last digit of a whole number shows how many *units* it contains, the second last digit gives the number of *tens* and so on. For example, the number twenty-four is written as:

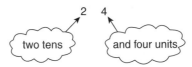

2 4

two tens   and four units,

However, when we start to use numbers which include bits of a whole (i.e. with decimals) some other 'places' are needed. These represent the tenths, hundredths, the thousandths, and so on. *The decimal point is simply a marker* to show where the units (whole numbers) end and the tenths begin. You'll get a better idea of this by discovering when the decimal point appears on your calculator. As you do Exercise 5.4, watch out for the decimal point …

Exercise 5.4 Blank checks

Complete the blanks and then check with your calculator.

100 ÷ 10 = ☐ ÷ 10 = ☐ ÷ 10 = ☐ ÷ 10 = ☐

400 ÷ 10 = ☐ ÷ 10 = ☐ ÷ 10 = ☐ ÷ 10 = ☐

25 ÷ 10 = ☐ ÷ 10 = ☐ ÷ 10 = ☐ ÷ 10 = ☐

1 ÷ 100 = ☐

4 ÷ 100 = ☐

7 ÷ 1000 = ☐

You may have got a picture of the decimal point jumping one place to the left every time you divide by 10. Actually, this is a slightly misleading picture. Most calculators work on a principle of a 'floating decimal point'. This means that the decimal point moves across the screen to keep its position between the units and the tenths digit.

The key point to remember is that the decimal point is nothing more than a mark separating the units from the tenths. Below I've written out the more complete set of 'place values' extending beyond hundreds, tens and units into decimals.

Hundreds	Tens	Units	Decimal point	Tenths	Hundredths	Thousandths
(100)	(10)	(1)	•	(0.1)	(0.01)	(0.001)

Using the four rules with decimals

If you aren't sure how to use the four rules of +, −, × and ÷ with decimal numbers, why don't you experiment with your calculator? You will quickly discover that the four rules work in exactly the same way for decimals as for whole numbers, and for that reason addition and subtraction of decimals are not spelled out here as a separate topic.

Back in Chapter 4, I suggested that you could learn about multiplying and dividing fractions by experimenting with your calculator. Exercise 5.5 is designed to help you do just that.

Exercise 5.5 Multiplying with decimal fractions

The first column in this table gives you three multiplication calculations involving fractions. For each calculation:

a change the fractions to decimals (column 2)

b use your calculator to multiply the decimals (column 3)

c change the decimal answer back to a fraction (column 4).

Calculation in fractions	Calculation in decimal form	Decimal answer (use a calculator)	Fraction answer
$\frac{1}{2} \times \frac{1}{2}$	0.5×0.5	0.25	$\frac{1}{4}$
$\frac{1}{2} \times \frac{1}{5}$			
$\frac{3}{5} \times \frac{1}{2}$			

Now look at columns 1 and 4 and see if you can spot the rule for multiplying fractions. Think about this for a while before reading on.

Spotlight: Rule for multiplying fractions

To multiply two fractions, say $\frac{3}{5}$ and $\frac{1}{2}$, multiply the numerators (3×1, giving 3), then multiply the denominators (5×2, giving 10).

The answer in this case is $\frac{3}{10}$: i.e. $\frac{3}{5} \times \frac{1}{2} = \frac{3 \times 1}{5 \times 2} = \frac{3}{10}$

Example 1

Look at the following multiplication:

$\frac{3}{4} \times \frac{2}{5}$

As before, these two fractions can be converted into decimal form, so the calculation can be rewritten as follows.

0.75×0.4

Pressing 0.75 $\boxed{\times}$ 0.4 $\boxed{=}$ on the calculator gives an answer of 0.3, or $\frac{3}{10}$.

By way of a check, we could apply the rule for multiplying fractions. We multiply the two numerators and then the two denominators, as follows.

$\frac{3}{4} \times \frac{2}{5} = \frac{3 \times 2}{4 \times 5} = \frac{6}{20}$

This can be simplified to $\frac{3}{10}$ (remember from Chapter 4 that $\frac{6}{20}$ and $\frac{3}{10}$ are equivalent fractions).

So, it doesn't matter whether multiplication is done in fraction form or decimal form; the result is the same either way.

Example 2

Look at this multiplication:

$2\frac{1}{4} \times 4\frac{3}{5}$

Again, these two fractions can be converted into decimal form, so the calculation can be rewritten as follows.

2.25×4.6

Pressing 2.25 $\boxed{\times}$ 4.6 $\boxed{=}$ on the calculator gives an answer of 10.35.

As before, we can check this against the rule for multiplying fractions. However, the number $2\frac{1}{4}$ must be rewritten as $\frac{9}{4}$, and $4\frac{3}{5}$ must be rewritten as $\frac{23}{5}$.

$\frac{9}{4} \times \frac{23}{5} = \frac{9 \times 23}{4 \times 5} = \frac{207}{20}$

Finally, just to check that the two methods produce the same result, this fraction can be converted to decimal form by dividing the numerator by the denominator.

Pressing 207 ⌈÷⌉ 20 ⌈=⌉ on the calculator confirms the previous answer of 10.35.

Dividing fractions

Suppose you want to find, say, $\frac{4}{5} \div \frac{1}{2}$. A rule of thumb for dividing fractions is to turn the fraction you are dividing by upside down and proceed as for multiplication. So the calculation becomes $\frac{4}{5} \times \frac{2}{1} = \frac{8}{5} = \frac{13}{5}$.

Note: a handy alternative strategy for dividing fractions is to convert the fractions to decimal form and perform the division on your calculator. Here are two examples in which the calculations have been done by both methods and, reassuringly, give the same answer in each case.

Example 3

$\frac{3}{4} \div \frac{2}{5}$

Rewriting as decimal fractions, this gives:

$0.75 \div 0.4$

Using the calculator, this gives an answer of 1.875.

Remember, one way of dividing fractions is to turn the fraction you are dividing by upside down and multiply. So, in this case we get:

$\frac{3}{4} \div \frac{2}{5} = \frac{3}{4} \times \frac{5}{2} = \frac{15}{8}$

If you check on your calculator (by pressing 15 ⌈÷⌉ 8 ⌈=⌉), you will see that the fraction $\frac{15}{8}$ has the same value as the earlier answer of 1.875.

A useful fact to remember about decimal numbers (and indeed about whole numbers) is that the value of each digit gets smaller as you go to the right. For example, in the number 0.235, the 3 has less value than the 2 since it refers to 3 hundredths, whereas the 2 is 2 tenths. A useful way of demonstrating this is to write the number as shown below.

0.235

The number is written so that the physical size of each digit indicates its unit value in comparison to the other units.

Example 4

Look at this division:

$4\frac{1}{5} \div 1\frac{1}{2}$

Again, these two fractions can be converted into decimal form, so the calculation can be rewritten as follows:

$4.2 \div 1.5$

Pressing 4.2 $\boxed{\div}$ 1.5 $\boxed{=}$ on the calculator gives an answer of 2.8.

As before, we can check this against the method of dividing fractions described above. However, the number $4\frac{1}{5}$ must be rewritten as $\frac{21}{5}$ and $1\frac{1}{2}$ must be rewritten as $\frac{3}{2}$.

$$\frac{21}{5} \div \frac{3}{2} = \frac{21}{5} \times \frac{2}{3} = \frac{21 \times 2}{5 \times 3} = \frac{42}{15}$$

which can be simplified to $\frac{14}{5}$.

Finally, just to check that the two methods produce the same result, this fraction can be converted to decimal form by dividing the numerator by the denominator, giving the same answer as before.

$14 \div 5 = 2.8$.

An overview of decimals

As your confidence with decimals grows, you will come to appreciate how decimal numbers are a natural extension of our whole number system. What this means in practice is being able to understand place value. So just as you add up to ten units and then swap them for one ten, so you add up to ten hundredths and swap them for one tenth.

Example 5

1	0.01
6	0.06
<u>3</u>	<u>0.03</u>
10	0.10

Ten units are written as 1 in the tens column.

Ten hundredths are written as 1 in the tenths column.

(Note that this result, 0.10, will be shown simply as 0.1 on the calculator display)

One obvious property of whole numbers is that the more digits a number has, the bigger it is. Unfortunately this is *not* true for decimal numbers. For example, the number 5.831659 is actually smaller than, say, 7.2. Don't be unduly impressed by a long string of digits. What matters is the position of the decimal point. You need to see beyond this string of digits and get a sense of how big the number actually is. For example, it is more useful to know that 5.831659 is between 5 and 6 (or just less than 6) than to quote it to six decimal places.

Sensibly used, calculators are an excellent means of seeing beyond the digits of a number. For example, earlier in the chapter, in Exercise 5.4, you were asked to perform repeated division by 10 and then observe what happened to the decimal point of the answer. This is an exercise which you can do with any starting number of your own choice, and the repeated division by 10 can be more efficiently done by using the calculator's constant facility.

Exercise 5.6 contains some calculator activities which should help you to become more confident with decimals.

Spotlight

A much debated mathematical fact is whether the number 0.9999... is equal to 1. Many students argue that this cannot be true, since 0.9999... clearly contains numbers that are 'just shy of 1', even though there is an infinite number of nines. However, here is an argument for showing why these two numbers are equal.

Write down $\frac{1}{3}$ as a decimal:

$$\frac{1}{3} = 0.3333...$$

Multiply both sides by 3:

left-hand side $= \frac{3}{3} = 1$
right-hand side $= 0.9999...$

So, it follows that 0.9999... = 1.

Exercise 5.6 Using the constant to investigate decimals

a Set your calculator's constant to divide by 10. Next enter a large number into the calculator display and then repeatedly press ⌐=⌐. At first, watch what happens. Later, try to predict what will happen.

b Set the constant to multiply by 10. Then enter a small decimal fraction and repeatedly press ⌐=⌐. Try to make sense of what is going on and then try to predict what will happen next.

c Set the constant to add 0.1 and repeatedly press ⌐=⌐. Without pressing the 'Clear' key, enter a large decimal number and keep pressing ⌐=⌐.

d Set the constant to add 0.01, and repeat what you have just done in part c.

e Repeat parts c and d but with the constant set for subtraction in each case.

f With a friend, play the game 'Guess the number', the rules of which are explained at the end of the chapter.

Practical situations involving decimals abound, the most obvious example being money. Thus £3.46 represents 3 whole pounds,

4 tenths of a pound (i.e. 4 ten-pences) and 6 hundredths of a pound (i.e. 6 pence).

However, the money representation of decimals can be confusing. We *say* £3.46 as 'three pounds forty-six', rather than 'three point four six', which is the more correct decimal form. This latter version emphasizes the decimal place value of each digit. Otherwise you can get into trouble when dealing with sums of money. For example, one pound and nine pence is often mistakenly written as £1.9, rather than £1.09.

We grow up with decimals and metric units, like metres, centimetres, kilograms, millilitres, and so on all around us. However, we still have feet and inches, pounds and ounces; these units, known as imperial units, are the ones that many adults still feel happy with. These units are explained in some detail in Chapter 7.

The main advantage of metric units is that they are based entirely on tens, hundreds and thousands. For example, there are 100 centimetres in a metre, 1000 metres in a kilometre, 1000 millilitres in a litre, and so on. Contrast this with the old-fashioned 14 pounds in a stone, 12 inches in a foot, 1760 yards in a mile, and so on – really a complete shambles!

Exercise 5.7 Practice exercise

1 **a** Mark the numbers 0.35 and 0.4 on the number line below.

```
|---|---|---|---|---|---|---|---|---|---|
0                      0.5                      1
```

   b Which of the two numbers, 0.35 or 0.4, is bigger?

2 In the number 0.6, the 6 stands for 6 _____.

3 Ring the number nearest in size to 0.78.

   0.7 70 0.8 80 .08 7

4 Multiply by 10: 5.49 → _____.

5 Add one tenth: 4.9 → _____.

6 The number marked with an arrow on the line below is about

_____.

21   22

7 How many different numbers can you write down between 0.26 and 0.27?

8 Which of these numbers is larger, 24.91257 or 83?

Guess the number

A game for two players, based on the calculator constant.

Player A secretly chooses a number between 1 and 20 – say 12 – and presses 1 ÷ 12 = 0. The final 0 is pressed in order to clear the display. (If your calculator has a 'double press' constant, then press 12 ÷ ÷ 0 instead.)

Player B has to guess which number *A* has chosen to hide in the calculator constant by trying different numbers and pressing =. The aim is for *B* to guess *A*'s number in the fewest possible guesses.

Sample play: *B*'s attempts to guess the hidden number 12 are as follows.

B presses	Display	Comments
16 =	1.3333333	16 is too big
15 =	1.25	15 is too big
9 =	0.75	9 is too small
12 =	1.	12 is the hidden number

Answers to exercises for Chapter 5

Your calculator should have provided you with most of the answers to these exercises. However, here are some of the main points.

EXERCISE 5.1

Key sequence	Fraction	Decimal	Conclusion
1 ÷ 2 =	$\frac{1}{2}$	0.5	
2 ÷ 4 =	$\frac{2}{4}$	0.5	
5 ÷ 10 =	$\frac{5}{10}$	0.5	
50 ÷ 100 =	$\frac{50}{100}$	0.5	The decimal for $\frac{1}{2}$ is ⟨0.5⟩
1 ÷ 4 =	$\frac{1}{4}$	0.25	
2 ÷ 8 =	$\frac{2}{8}$	0.25	
5 ÷ 20 =	$\frac{5}{20}$	0.25	
25 ÷ 100 =	$\frac{25}{100}$	0.25	The decimal for $\frac{1}{4}$ is ⟨0.25⟩
3 ÷ 4 =	$\frac{3}{4}$	0.75	
6 ÷ 8 =	$\frac{6}{8}$	0.75	
75 ÷ 100 =	$\frac{75}{100}$	0.75	The decimal for $\frac{3}{4}$ is ⟨0.75⟩
1 ÷ 10 =	$\frac{1}{10}$	0.1	
10 ÷ 100 =	$\frac{10}{100}$	0.1	The decimal for $\frac{1}{10}$ is ⟨0.1⟩

EXERCISE 5.2

Fraction	$\frac{1}{2}$	$\frac{1}{4}$	$\frac{3}{4}$	$\frac{1}{10}$	$\frac{1}{5}$	$\frac{2}{5}$	$\frac{3}{10}$	$\frac{9}{10}$	$\frac{1}{20}$	$\frac{1}{8}$	$\frac{1}{3}$
Decimal	0.5	0.25	0.75	0.1	0.2	0.4	0.3	0.9	0.05	0.125	0.33...

EXERCISE 5.3

Calculation	Press
0.5 $\boxed{+}$ 0.5 $\boxed{=}$	1.
0.5 $\boxed{\times}$ 2 $\boxed{=}$	1.
0.25 $\boxed{\times}$ 4 $\boxed{=}$	1.
0.5 $\boxed{\times}$ 10 $\boxed{=}$	5.
4 $\boxed{\div}$ 10 $\boxed{=}$	0.4
0.1 $\boxed{+}$ 0.1 $\boxed{=}$	0.2
0.1 $\boxed{\times}$ 10 $\boxed{=}$	1.

EXERCISE 5.4

$100 \boxed{\div} 10 \boxed{=} 10 \boxed{\div} 10 \boxed{=} 1 \boxed{\div} 10 \boxed{=} 0.1 \boxed{\div} 10 \boxed{=} 0.01$

$400 \boxed{\div} 10 \boxed{=} 40 \boxed{\div} 10 \boxed{=} 4 \boxed{\div} 10 \boxed{=} 0.4 \boxed{\div} 10 \boxed{=} 0.04$

$25 \boxed{\div} 10 \boxed{=} 2.5 \boxed{\div} 10 \boxed{=} 0.25 \boxed{\div} 10 \boxed{=} 0.025 \boxed{\div} 10 \boxed{=}$
0.0025

$1 \boxed{\div} 100 \boxed{=} 0.01$
$4 \boxed{\div} 100 \boxed{=} 0.04$
$7 \boxed{\div} 1000 \boxed{=} 0.007$

EXERCISE 5.5

Calculation in fractions	Calculation in decimal form	Decimal answer (use a calculator)	Fraction answer
$\frac{1}{2} \times \frac{1}{2}$	0.5×0.5	0.25	$\frac{1}{4}$
$\frac{1}{2} \times \frac{1}{5}$	0.5×0.2	0.1	$\frac{1}{10}$
$\frac{3}{5} \times \frac{1}{2}$	0.6×0.5	0.3	$\frac{3}{10}$

EXERCISE 5.6
No comments.

EXERCISE 5.7

1 a

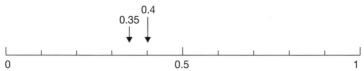

b 0.4 is bigger than 0.35

2 In the number 0.6, the 6 stands for 6 tenths.

3 The number nearest in size to 0.78 is ringed below.
0.7 70 (0.8) 80 .08 7

4 5.49 → 54.9

5 4.9 → 5.0

6 The arrowed number is about 21.85.

7 There are infinitely many numbers between 0.26 and 0.27.
For example, I could write out the thousandths: 0.261, 0.262,
0.263, and so on up to 0.269. There are nine of these. But
between, say, 0.262 and 0.263 I could write out nine further
numbers, each expressed as a ten thousandth: 0.2621, 0.2622,
0.2623, and so on. Then I can write out numbers in hundred
thousandths, millionths, and so on. This process can continue
indefinitely or until I fall over with exhaustion.

8 Although 24.91257 contains more digits than 83, its value is
only about 25, so 83 is larger.

Summary

This chapter should have helped you to make the link between fractions and decimals. I hope that after reading it, you now have a clearer sense of how decimal fractions (i.e. numbers like 0.56, 45.03 and so on) fit into the way the number system is organized. While digits to the left of the decimal point represent the number of units, tens, hundreds, and so on, the digits to the right are the tenths, hundredths, thousandths, and so on.

If you want to explore numbers, the calculator is an excellent place to start. A variety of calculator activities were suggested which should all contribute to your understanding of, and confidence with, decimals.

Finally, here is a checklist of the sort of things you should aim to know about decimals. *Note*: You will have the opportunity of using decimals again when we look at units of measure in Chapter 7.

CHECKLIST

You should have the ability to:

▶ know that the 4 in the number 6.143 refers to four hundredths

▶ mark decimal numbers on the number line

▶ arrange decimal numbers in order from smallest to biggest

▶ multiply and divide decimal numbers by 10, 100 and 1000

▶ know that 3.45 is half way between 3.4 and 3.5

▶ know that 0.25 means $\frac{1}{4}$ and 3.75 means $3\frac{3}{4}$

▶ handle units (metres, pounds (£), kilograms) in practical situations

▶ know roughly what answer to expect in a calculation involving decimals.

To find out more via a series of videos, please download our free app, *Teach Yourself Library*, from the App Store or Google Play.

6

Percentages

In this chapter you will learn:

▶ *why percentages are important*

▶ *about the connection between percentages, fractions and decimals*

▶ *how to do percentage calculations*

▶ *about common difficulties that people experience with percentages.*

The facts are right but the conclusion is wrong when you realize that the population of China is about 300 times that of New Zealand. In fact, in New Zealand there is roughly one car for every two people – in China the corresponding figure is roughly one car for every 14 people.

As we can see, failing to compare like with like can lead to incorrect conclusions. Percentages are useful for making fair comparisons, but unfortunately many find them difficult.

Government reports and educational research has confirmed that many adults don't understand percentages. This is despite the fact that newspapers and TV use the words 'percentage' and 'per cent' over and over. I opened a daily newspaper at random and quickly picked out these examples.

Read the cuttings below and try to make sense of how the term 'per cent' is being used. You can read them again at the end of the chapter.

Babies behind bars

Women make up just 5% of the prison population and 60% are serving sentences of six months or less. But because there are fewer women's prisons, they are on average 55 miles from home, far further than for men.... Women try to do everything they did outside prison – worrying if the kids are okay, trying to sort out bills and solve problems. Male prisoners largely rely on women to maintain domestic stability, but of around 20 000 children with mothers inside, only 9% are cared for by their fathers.

Source: *Radio Times*, 21 May 2016, page 23, by Janice Turner

Smoking in figures

About 5.8 trillion (5 800 000 000 000) cigarettes were smoked worldwide in 2014. Between 1990 and 2009, cigarette consumption fell by 26% in Western Europe but increased by 57% in Africa and the Middle East.

Nearly one in five (19%) UK adults were smokers in 2013, down from 26% in 2003 and the lowest level since records began in 1982.

In 2013, 12% of UK adults in managerial/professional jobs smoked, compared with 29% of those in manual occupations.

Source: *The Observer*, 22 May 2016, main section, page 15, special report by Jamie Doward

The reason why most people are confused by percentages is, I think, quite simple. Many adults and most children don't really understand what a percentage is.

What is a percentage?

A percentage is very similar to a decimal and a fraction. Like them, it describes a 'bit' of a number. It is really a particular sort of fraction.

Think back to Chapter 4, which explained how fractions could be *equivalent*. Here are some equivalent fractions:

$$\frac{1}{2}, \frac{2}{4}, \frac{5}{10}, \frac{50}{100}$$

They are equivalent because they all have the same value of a half.

Now look at the last of these four fractions, $\frac{50}{100}$. You might read it as 'fifty out of a hundred'. A shorthand way of saying this uses the Latin words *per centum* meaning 'out of every hundred'.

So, $\frac{50}{100}$ is the same thing as 'fifty per cent'.

out of   a hundred

Well, if a half (i.e. $\frac{50}{100}$) is the same as 50 per cent, what do you think a quarter becomes as a percentage?

The answer is 25 because $\frac{1}{4} = \frac{25}{100}$ or 25 per cent.

The symbol for 'per cent' is %.

So 25% is really another way of writing $\frac{25}{100}$ or 25 per cent.

Spotlight

The symbol for per cent is %, which gives a clue to its meaning. The two zeros signal that the number 100 is involved; while the fact that these digits are above and below a quotient line, suggest that it is a fraction. Put these two clues together and you've worked out that a percentage is a fraction 'out of 100'. In other words, it is equivalent to a fraction with a denominator (the number on the bottom) as 100.

Changing a fraction to a percentage

By now you might have worked out for yourself how to change a fraction to a percentage. If not, you can read the method which I've summarized in two simple steps below. Let us take the example of converting the fraction $\frac{4}{5}$ to a percentage.

Step 1   Change the fraction into its decimal form, so $\frac{4}{5} \rightarrow 0.8$

(If you find this hard to do in your head, press 4 $\boxed{\div}$ 5 $\boxed{=}$ on your calculator.)

Step 2   Express the answer in hundredths.
0.8 is the same as 0.80, or 80 hundredths.
So, $\frac{4}{5}$ converts to 80%.

You will probably want to practise this, so try Exercise 6.1 now.

Exercise 6.1 Changing fractions to percentages

Fill in the blanks in the table below. The first one has been done for you.

Fraction	Decimal fraction	Percentage
$\frac{1}{2}$	0.5	50%
$\frac{3}{4}$		
$\frac{7}{10}$		
$\frac{1}{5}$		
$\frac{1}{20}$		
$\frac{3}{5}$		
$\frac{3}{8}$		

Like fractions and decimals, percentages can be represented on the number line. In the percentage number line below, 100% corresponds to the number 1, 200% to 2 and so on.

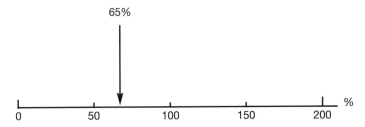

65%

0   50   100   150   200   %

In practice, percentages are rarely represented in the form of a number line but I've included it to stress the similarity with fractions and decimals.

Why bother with percentages?

The main advantage of percentages is that they are much easier to compare than fractions. For example, which do you think is bigger, $\frac{3}{4}$ or $\frac{7}{10}$? Written like this you can't really say, because the slices of the whole 'cake' (quarters and tenths, respectively) are not the same size. In order to make a proper comparison, the fractions need to be broken down to the same size of slice, and hundredths are very convenient. So here goes…

Fraction	Hundredths	Percentage
$\frac{3}{4}$	$\frac{75}{100}$	75%
$\frac{7}{10}$	$\frac{70}{100}$	70%

Clearly 75% is bigger than 70%, so we can now conclude that $\frac{3}{4}$ is bigger than $\frac{7}{10}$.

Spotlight

Did you know that newborn babies, both male and female, are composed of about 78% water? And that, by maturity, males are about 60% water and females 55%?

If you look at a practical example you will get a better idea of how useful percentages are.

Which of the following would represent the bigger price rise?

a Bread is to go up by 6p per loaf.

b A camera is to go up by £5.

In one sense the answer could be **b,** because £5 is more than 6p. But, since most people buy many more loaves of bread than they do cameras, we would probably be more concerned if bread went up by 6p per loaf. The only fair way to compare these price rises is to acknowledge that 6p is a lot compared with the price of a loaf of bread, whereas £5 may not be so much compared with the price of a camera. Using percentages allows us to make comparisons, taking account of the prices of each item.

So, if we convert these price rises to percentage price rises, a very different picture emerges. In Exercise 6.2 you are asked to have a go at calculating these two percentage increases. Don't worry if you can't do it straight away, as the method is explained below.

Exercise 6.2 Calculating percentage increases

Complete the table below. (I've taken the original price of bread to be 60p per loaf and that of the camera to be £100.)

	Original price (£)	Price rise (£)	Percentage price rise
Bread	0.60	0.06	
Camera	100.00	5.00	

SOLUTION

The calculation of the percentage price increases is illustrated as follows.

	Original price (£)	Price rise (£)	Percentage price rise
Bread	0.60	0.06	$\frac{0.06}{0.60} \times 100 = 10\%$
Camera	100.00	5.00	$\frac{5}{100} \times 100 = 5\%$

In summary, then, percentage price increases (or decreases) are calculated as follows.

$$\text{Percentage price increase} = \frac{\text{Price increase}}{\text{Original price}} \times 100$$

Perhaps you were able to confirm my calculation that bread went up in price by 10 per cent, whereas the camera went up by only 5 per cent in price.

There is, of course, another reason that this increase in the price of bread will cause more concern than that of the camera. It is that we tend to buy bread every week, so this price rise is affecting our shopping bill every week. Cameras, on the other hand, are a very rare purchase, and even a £5 price rise will simply not affect most people most of the time.

Let's now look in more detail at how to calculate percentage increases and reductions.

Spotlight: Can you turn, say, the number 43 into a percentage?

The answer is no! You cannot turn a single number into a percentage. It must be expressed in comparison to (i.e. 'out of') some other number. Thus, 43 out of 50 can be turned into a percentage – it is $\frac{43}{50}$ or 86%.

Calculating percentage increases and reductions

Where the percentages convert to very simple fractions (e.g. 100 per cent, 50 per cent, 25 per cent or 10 per cent), it should be possible to do the calculation in your head. However, for anything more complicated, I would always use a calculator to calculate percentage changes. Here, first, are two examples which could probably be done in your head.

Example 1 A 50% reduction

SOCKS
£~~2.50~~ a pair
Now 50% off!!

What is the sale price of a pair of these socks?

Solution

Since 50% is $\frac{1}{2}$, there is a reduction of half of £2.50.

This reduction is $\frac{£2.50}{2} = £1.25$.

So the new price is £2.50 – £1.25 = £1.25.

Example 2 A 5% increase

In 2008, the average price of a new house in a particular town in the Midlands was £182 000.

Over the next year, prices of new houses in the town increased by about 5%.

Estimate the average price of a new house a year later.

Solution

Since 5% is the same as $\frac{1}{20}$, we know that the prices rose by $\frac{1}{20}$ over this period.

So, the price rise is $\frac{182\,000}{20} = £9100$.

Adding to the original price, we get:

An estimate of the average price of a new house a year later is £182 000 + £9 100 = £191 100.

So much for calculating simple percentage increases using pencil and paper only. Unfortunately, most percentage calculations are more complicated than this and require a calculator.

The method for calculating percentage price changes is explained in the next two examples.

Example 3 A 6% increase

Following a budget announcement on petrol tax, garages increased all their pump prices by 6%.

> Current petrol prices
>
> Unleaded premium, 97.2p per litre.
>
> All prices to go up by 6% at midnight tonight.

What is the new price of unleaded premium petrol at this garage?

Solution

An increase of 6% means an increase of six hundredths, or in other words, an increase of 0.06 of the original price. A possible way of proceeding here is to perform the calculation in two stages. First, find the price increase and then add it on to the original price. As you will see shortly, there is a quicker, one-staged method, but you will find this method easier to follow, if you first work through the two stages explained below.

Stage 1: Find the price increase.

The price increase is $0.06 \times 97.2p = 5.832p$.

Petrol prices are usually quoted to one decimal place, so this price increase would be *rounded* to the nearest tenth of a penny, i.e. 5.8p.

Stage 2: Add on the price increase.

The new price is $97.2p + 5.8p = 103.0p$.

(Notice that the price is written as 103.0p, rather than 103p, in order to stress that the price has been stated accurate to one decimal place.)

As was suggested above, if all you want to find is the new price, this two-staged method is unnecessarily complex. The whole process can be reduced to a single stage, by multiplying the old price by 1.06.

It may not be obvious to you where the 1.06 comes from. It is helpful here to think in terms of hundredths. Before the 6% price increase we have $\frac{100}{100}$ of the given amount. Adding 6% will increase this to $\frac{100}{100} + \frac{6}{100} = \frac{106}{100}$, which equals 1.06.

So, the new price is $1.06 \times 97.2 = 103.032$p.

This rounds to 103.0p, which confirms the previous answer from the two-staged method.

Example 4 A 15% reduction

This time the socks sale is rather less inviting. As you can see, the reduction now is only 15%!

> # SOCKS
> £2.50 a pair
> ## Now 15% off!!

What is the sale price of a pair of these socks?

Solution

Again, since 15% is not easily converted into a convenient fraction, it makes sense to do this calculation on a calculator. As for Example 3, I will first do it the long-winded, two-staged way and then more directly using the one-staged method.

Stage 1: Find the price decrease.

The price decrease is $0.15 \times 2.50 = £0.375$.

Stage 2: Subtract the price reduction.

The new price is $£2.50 - £0.375 = £2.125$.

Rounding to the nearest penny, the new price is £2.13.

As was the case with Example 3, the whole process can be reduced to a single stage, by multiplying the old price by 0.85. Again, it is helpful to think in terms of hundredths. Before the 15% price decrease we have $\frac{100}{100}$ of the given amount. Subtracting 15% will decrease this to $\frac{100}{100} - \frac{15}{100} = \frac{85}{100}$ which equals 0.85.

So, the new price is $0.85 \times £2.50 = £2.125$.

After rounding, this confirms the previous answer from the two-staged method.

You will need some practice at calculating percentage increases and decreases, so have a go at Exercise 6.3 now.

Exercise 6.3 Calculating percentage increases and decreases

1 A chair normally sells at £42. How much will it cost with a 30% reduction?

2 Is my garage bill correct? If not, what should the correct bill be?

Anytown Autos
Full service repairs and parts	£186.40
VAT @20%	£39.28
Total (inc. VAT)	£233.68

3 If you earn £230 per week, which would you prefer: a rise of 6%, or a rise of £12 per week?

4 Your taxable earnings are £884 this month. How much of this will you have left after paying 33% in stoppages?

Spotlight

The managing director of a particular company was being interviewed and claimed that: 'Our profit increased this year by 70% over last year.'

Are you impressed?

The problem here is that he has not stated the sort of sums of money involved. If, over one year, a large company were to increase its profit from, say, £100m to £170m – an increase of 70% – then you would be right to be impressed. But suppose his company barely broke even over the previous year – let's say they cleared just £10 profit. They would only have to make a profit this year of £17 to show a 70% increase in their profit. Big deal!

Persistent problems with percentages

It will be no surprise to you to be told that a lot of children's time in school takes place with one eye shut and the other staring out of the window. Many children come away from a lesson in percentages (or whatever) with only a few pieces of the jigsaw and have to somehow fill in the rest of the picture themselves. Unfortunately they don't always get it right. Can you spot where the following child has gone wrong?

The problem is that she has started from a true fact that $10\% = \frac{1}{10}$ and built up a rule which doesn't work for any other fraction. This is probably the most common misapprehension about percentages. If you still have problems with this, the chances are that they can be traced back to a fuzziness about fractions. You may know that 20% is more than 5%. However, it is *not* the case that $\frac{1}{20}$ is more than $\frac{1}{5}$. If you think back to Chapter 4 and the idea of a fraction being a slice of cake, then imagine a cake cut into twenty equal slices. Each of these slices is a twentieth of the cake and is therefore a very small slice indeed. One fifth, on the other hand is a large slice.

$\frac{1}{20}$ = 5%

$\frac{1}{5}$ = 20%

Shopping during the sales is an opportunity to check out some of these ideas. For example, $\frac{1}{3}$ off is a better discount than 10% off. Also remember that 40% off the price of something fairly cheap (like a packet of envelopes) represents only a small saving in actual money, whereas the same percentage reduction from, say, the price of a house represents a huge saving.

Perhaps the most important thing you need to grasp is that fractions, decimals and percentages are really the same thing. I have found that drawing three number lines one above the other is a helpful way of emphasizing these connections, as shown below. The arrows show that $\frac{3}{5}$, 0.6 and 60% have the same value.

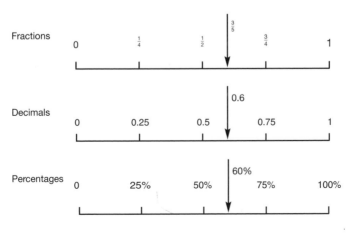

Exercise 6.4 Practice exercise

1 Which is bigger, 8% or $\frac{1}{8}$?

2 Which is bigger, 15% or $\frac{1}{15}$?

3 If the rate of inflation drops from 5% to 4%, are prices:
   a going up? **b** coming down? **c** neither?

4 What is 20% of £80?

5 What is 10% of 20% of £80?

6 Here are some egg prices before and after a price rise.

	Old price (per half dozen)	New price (per half dozen)
Small eggs	71	75
Large eggs	84	88

Which has had the greater price increase: small or large eggs?

7 My garage bill has come to £120.93 and includes VAT at 20%. What would the bill be:

   a without VAT?

   b if VAT were rated at 25% instead of 20%?

8 Why do children's sweets tend to suffer greater inflation than most of the other things adults buy?

9 Study the newspaper cuttings given at the beginning of this chapter and then answer the following questions.

   a From the first cutting, how many children with mothers in jail are cared for by their fathers? Why might your answer not be an exact figure?

   b From the second cutting, how have smoking habits changed globally in recent years?

Answers to exercises for Chapter 6

EXERCISE 6.1

Fraction	Decimal fraction	Percentage
$\frac{1}{2}$	0.5	50%
$\frac{3}{4}$	0.75	75%
$\frac{7}{10}$	0.7	70%
$\frac{1}{5}$	0.2	20%
$\frac{1}{20}$	0.05	5%
$\frac{3}{5}$	0.6	60%
$\frac{3}{8}$	0.375	$37\frac{1}{2}\%$

EXERCISE 6.2

The solution is in the text.

EXERCISE 6.3

1 The reduction in price is 30% or $\frac{3}{10}$. Three tenths of £42 can be found on your calculator by pressing

either

3 ÷ 10 × 42 =

or

0.3 × 42 =

both of which give the correct answer £12.60.

So the reduced price is £42 – £12.60 or £29.40.

(*Note*: A quicker way of doing this is to say that a 30% reduction will bring the price down to 70% of the old price. On the calculator you would press

0.7 × 42 =, which gives the answer £29.40.)

2 The VAT is incorrect. Using a calculator, press 186.40 × 0.2 = to give the answer £37.28. The addition was also incorrect, so I've been overcharged by £10. The correct total is £223.68.

(*Note*: the direct method for checking the final bill is to press 186.40 ☒ 1.2 ☐.)

3 6% of £230 is £13.80, which is a bigger rise than £12.

4 I will have 67% of £884 left, which is £592.28.

EXERCISE 6.4

1 $\frac{1}{8}$ is $12\frac{1}{2}$% and so is bigger than 8%.

   Note: You can use your calculator to convert $\frac{1}{8}$ to a percentage by pressing 1 ☐ 8 ☒ 100 ☐.

2 $\frac{1}{15}$ is approximately equal to 6.7%, so 15% is bigger than $\frac{1}{15}$.

3 The annual rate of inflation measures how much average prices have risen over a year. If that rate is a positive number (such as 4% or 5%, for example), this means that prices have risen. So, even though the rate of inflation has fallen, the current rate of 4% shows that prices are still rising, but not quite as quickly as they were over the previous year.

4 20% is $\frac{1}{5}$. One fifth of £80 is $\frac{£80}{5} = £16$.

5 10% of 20% is one tenth of 20%, which is 2%.
   2% of £80 = £80 $\times \frac{2}{100}$ = £1.60

6 The solution is summarized in the table below.

	Old price (p)	New price (p)	Price increase (p)	Percentage increase (rounded to 1 decimal place)
Small eggs	71	75	4p	$\frac{4}{71} \times 100 = 5.6\%$
Large eggs	84	88	4p	$\frac{4}{84} \times 100 = 4.8\%$

So, although both eggs have seen the same actual price rise (4p in each case), the small eggs have shown the greater percentage rise.

7 **a** This calculation is slightly harder than the others, as it involves working backwards after the percentage increase has been added. The story line of the solution is as follows: Let the bill without VAT be thought of as 100% and the bill with VAT (costing £120.93) as 120%. So we must divide the total bill by 120 and then multiply by 100.

Now, $120.93 \times \frac{100}{120} = 100.78$ (rounded to the nearest penny). So the net bill (i.e. not including the VAT) is £100.78.

  b To calculate the bill inclusive of VAT at 25%, multiply the net bill by 1.25.

  £100.78 × 1.25 = £125.98.

8 When percentage increases are applied to the price of goods, it is generally the case that fractions of pence get rounded up. For example, if a bar of chocolate costing 36p is subjected to a 10% increase, the true price should be 39.6p. However, as shopkeepers cannot charge 0.6 of a penny, this price is likely to be rounded up to 40p. This represents a loss of 0.4p to the customer and a loss of 0.4p represents a much greater proportion of something costing 40p than of an item costing, say, £40. Since the things that children buy (comics, sweets, etc.) tend to be cheap, children lose out from these rounding losses more than adults.

9 a There are 20 000 children with mothers in prison, of whom 9% are looked after by their fathers. To calculate 9% of 20 000, multiply 20 000 by the fraction $\frac{9}{100}$. This is $20\,000 \times \dfrac{9}{100}$, or:

  $$\dfrac{20\,000 \times 9}{100}$$

  You could use a calculator here but it's not hard to do it in your head. First cancel out 100, which leaves 200 × 9. Multiplying out gives the answer, 1800 children.

  This will not be an exact answer for two main reasons. Firstly, the figures of 9% and 20 000 may not be accurate – for example, the figure 20 000 is too neat to be believable as exactly that value. It is likely that this is a rounded figure (the actual figure might be something like 20 814 or 19 102, for instance). Secondly, it has not been stated when the data were collected. As these sorts of statistics change constantly, the figure today will almost certainly be slightly different from the one you calculated.

  b To answer this question, you need to run your eye down each of the key facts and decide which one helps to answer your question. In this example, it is the second point that is the most relevant. It suggests that cigarette consumption is falling in Western Europe but greatly increasing in Africa and the Middle East. Since the combined populations of Africa and the Middle East exceed that of Western Europe and also the figure of 57% is much greater than 26%, this suggests that, globally, cigarette consumption is increasing. In fact, today there are roughly 1.1 billion tobacco users and this figure is expected to rise to 1.6 billion over the next 20 years.

Summary

You should now be able to:

▶ realize that converting to percentages makes it easier to compare fractions

▶ link percentages to common and decimal fractions

▶ convert from percentages to (simple) fractions and decimals and vice versa, e.g. $75\% = \frac{3}{4} = 0.75$

▶ express something as a percentage of something else, e.g. 6 is 25% of 24

▶ calculate percentage increases and decreases.

To find out more via a series of videos, please download our free app, *Teach Yourself Library*, from the App Store or Google Play.

7

Measuring

In this chapter you will learn:

▶ *about measuring dimensions and units*
▶ *how to round numbers*
▶ *how to convert units of measure.*

> **It is said of frogs that they sort all other animals they meet into just three categories.**
>
> If it is small, they eat it.
>
> If it is large, they run away from it.
>
> And if it is about their own size, they mate with it.

I think it's fair to say that, in general, humans are slightly more discriminating! Any activity which involves making judgements about the size of things can be called measuring. Although frogs may not be engaged in highly sophisticated measuring here, they *are* trying to understand the bigness or smallness of things around them.

What do we measure?

Most people tend to think of measuring as using weighing scales or a tape measure. But what sort of things do these devices tell us about? Weighing scales tell us about weight, and a tape measure about length. These types of measurement are called *dimensions*.

Dimensions of measure are measured in certain measuring *units*. For example, weight may be measured in kilograms, grams, ounces, pounds, and so on; while length may be measured in centimetres, metres, inches, miles, and so on.

There are, of course, many dimensions other than length and weight which we need to measure, for example:

temperature, time, area, capacity, angle, volume, speed ...

Exercise 7.1 will give you a chance to think about these dimensions and also about the units in which they are usually measured.

Exercise 7.1 Dimensions and units

Complete the table (the first two have been done for you).

Question	Dimension of measure	Likely units of measure
How heavy is your laundry?	weight	kg or lb
How long is the curtain rail?	length	cm or in
How hot is the oven?		

How far is it to London?

How fast can you run?

How long does it take to cook?

How much does the jug hold?

How big is your kitchen?

How big is the field?

As you can see, I have included two lots of units in the examples above, because both are in common usage. They are known as the metric system of units and the imperial system of units. Because many people are confused by these various units of measure, they are explained in some detail later in the chapter.

Spotlight: Start at zero

Suppose you are asked to estimate the time taken for some short event to take place – say the interval, in seconds, between a lightning flash and the ensuing rumble of thunder. If you start counting 1, 2, 3 ... you are actually chopping off the first second. As the diagram below shows, you should start counting from zero.

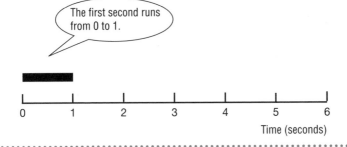

The first second runs from 0 to 1.

0   1   2   3   4   5   6

Time (seconds)

Why do we measure?

The reason we measure is that, quite simply, we live in a more complex world than a frog. Although words like 'large' and 'small' are sometimes good enough for some particular purposes ('Give me some of the *large* apples', 'I'd like a *small* helping', and so on), often we need to be more precise. Here is an example where the word 'large' proved inadequate during a national rail strike.

The use of the word 'large' in these quotations is highly dubious. How large would the proportion have to be for you to consider it 'large' – 10%, 25%, or perhaps 70%? … It is interesting that both sides in the dispute have been deliberately vague about the exact figures, and prefer instead to give a general impression.

Sometimes, however, a general impression is simply not good enough, and something more precise is needed. For example, you may have seen signs on the motorway advising drivers of 'large' vehicles to stop at the next emergency phone and contact the police. If you are driving a lorry, how would you know whether this referred to you? Rest assured that the small print below the sign goes on to explain that:

> **Large means 11' 00" (3.3 m) wide or over**

The reason that we tend to measure with numbers is to help us make decisions and comparisons fairly and accurately. Careful measuring helps us bake 'the perfect cake' every time, lay well-fitting carpets with a minimum of waste, check that the children's shoes don't pinch and so on. In Exercise 7.2 you are asked to think about the measuring dimensions involved in these sorts of everyday tasks.

Exercise 7.2 Being aware of the dimensions of measure

Here is a list of eight common dimensions of measure:

Length (L), Area (A), Volume (V), Weight (W), Time (T), Temperature (T°), Capacity (C) and Speed (S).

Make a note of which of them are likely to be important in the following everyday activities. (*Note*: there may be several dimensions involved in each activity.)

Everyday activities

▶ Baking a cake

▶ Buying and laying a carpet

▶ Checking the children's shoes

▶ Setting out on a journey in good time

Spotlight: Whose foot?

Early measures of length were based on using parts of the body as standard units. This approach gave rise to:

▶ the inch (the thickness of a man's thumb)

▶ the foot (well, this one is obvious)

▶ the yard (the length of a man's stride)

One problem with this approach was standardization of the units – not everyone's thumb, foot, stride, etc. are the same size.

How do we measure?

Measuring is a way of describing things. Descriptions can come in two basic forms. First, there are descriptions of *quality* and these tend to be made with words. For example, you may describe your various friends as happy, carefree, moody, thoughtful, sensitive, and so on. These are not the sorts of descriptions that easily lend themselves to being reduced to numbers. Descriptions of *quantity*, on the other hand, do involve numbers. For example, Ann is 1.59 m tall, Donal is 73 years old, Chris has 5 children, and so on. When people talk about measurement, they are usually thinking about measurement based on numbers, but not always.

Some measuring seems to fall between quality and quantity. For example, you might describe something as being large or small, fast or slow, chilly or warm. These are ways of indicating whereabouts on some sort of scale (respectively they refer to size, speed and temperature). Yet, although they make no mention of numbers, these descriptions are a sort of measure. Such words can be *ranked* into a meaningful order and they then produce what is called an *ordering* scale. On the other hand, words which describe, say, an emotion or a colour do not normally relate to a useful scale. Thus, you can't say that 'curious' is bigger than 'excited', or that 'red' is more than 'blue'. Such words are simply descriptions. Exercise 7.3 will give you practice at using an ordering scale.

Exercise 7.3 Using an ordering scale

Here are five words used in describing how *likely* something is to happen:

likely, impossible, doubtful, certain, highly improbable

These descriptive words can be written in order of likelihood, from least likely to most likely, thus:

impossible, highly improbable, doubtful, likely, certain

Now, here are some for you to do.

Rank the following sets of words into useful ordering scales.

1 These five words are used in describing ways of travelling on foot:

jog, stop, sprint, walk, amble

2 These are the developmental stages that babies usually go through:

walk, lie, sit up, stand, roll over

3 These words are often written in sequence on an electric iron:

wool, linen, silk, cotton, rayon

To summarize, then, measuring can take the following three basic forms:

▶ words alone

▶ words which can be ranked in order

▶ numbers.

The types of measuring scale which these three approaches use are:

▶ words

▶ ordering scale

▶ number scale.

Although all three types of scale are helpful in providing an interesting variety of descriptions and comparisons, it is the third of these, measuring with numbers, which is the most important in mathematics.

Spotlight

Until the early 1990s, temperatures in the UK (for example, weather and cooking temperatures) tended to be given using the Fahrenheit scale. This scale was named after Daniel Fahrenheit (1686–1736), who proposed it in 1724. On the Fahrenheit scale, the freezing point of water is 32 degrees Fahrenheit (°F) and the boiling point is 212°F (both measured at standard atmospheric pressure). This gives a scale of 212 − 32 = 180 units.

Fahrenheit's reason for choosing this odd starting point of 32°
to represent the freezing point of water, is to do with the coldest temperature that he was ever able to measure in his home town of Danzig (now Gdansk in Poland), which he defined as zero degrees. Based on his scale, the freezing point of water turned out to be 32 degrees warmer than this. Today, this scale has largely been replaced by the Celsius scale. On the Celsius scale, the freezing point of water is 0 degrees Celsius (0°C) and the boiling point is 100°C. This gives a scale of 100 − 0 = 100 units, which gives rise to its alternative name, the Centigrade (literally, 100 unit) scale.

How accurately should we measure?

The accuracy with which we measure depends entirely on what and why we are measuring. A brain surgeon and a tree surgeon have different needs for accuracy when sawing up their respective 'patients'. A nurse weighing out drugs will exercise greater care and precision than a greengrocer weighing out potatoes. A calculator can sometimes give a false sense of the accuracy of an answer. As the example below shows, it may give a result showing eight-figure accuracy, but the numbers on which the calculation was performed may be only approximate.

Suppose you wish to replace the fence in your garden. The length of fencing needed is, say, 21 m and each panel of fencing is 1.8 m in length. Pressing 21 ÷ 1.8 = on your calculator will probably produce the answer 11.666666 (for reasons that will be explained shortly, on some calculators the answer will be shown as 11.666667).

For this sort of calculation, it is plainly silly to give an answer to eight figures. If the last six digits of your answer are either dubious or unnecessary, then dispose of them. However, you have to be a bit careful how you do this. Numbers can be shortened so that you finish with a suitable number of digits (say three). This is called giving your answer 'correct to three significant figures' (or 'to 3 sig. figs.', for short). With the above example, calculating the number of panels of fencing requires that you buy a whole number of panels, so the answer will be given correct to two significant figures. In this case 11.666666 would be rounded up to 12 panels. *Note*: You would still need 12 panels even if the answer on the calculator was 11.333333!

This process of simplifying unnecessarily accurate measurements to a near approximation is called *rounding*. Some examples are given in the table below.

Measurement	Rounded to 3 sig. figs.
4.18345926	4.18
371.41429	371
0.0142419	0.0142
74312.692	74300
11.6666	11.7

Notice that the third and fourth examples in this table have produced answers which contain not three but five figures. However, the two zeros at the beginning of 0.0142 and the two zeros at the end of 74 300 are not considered to be significant figures. They are only there to give the overall magnitude of the number. It makes sense to do this, as otherwise the number 74 312.692 would be rounded to 743, which is clearly nonsense!

The last example in the above table, 11.6666, is different from the others in the following respect. As you can see, its third digit has been *rounded up* from a 6 to a 7. The clue to why it has been rounded up can be found by looking at the fourth digit in the original number: the 6. Since it is bigger than 5, the 6 in the tenths column is *rounded up* to a 7. And this is the reason that some calculators produce the answer 11.666667 for the fence panel calculation. Such calculators have been designed so that they round up the final digit displayed, when the next digit would have been a 5 or greater. In order to be able to do this, these calculators need to process their calculations to greater accuracy than the eight figures that they display, which accounts for why they tend to be slightly more expensive than the calculators which don't round. Exercise 7.4 gives you practice at rounding.

By the way, don't worry if this explanation of rounding sounds confusing – it is easier to do than to read about!

Exercise 7.4 Rounding practice

Round the following numbers to four significant figures.

	Number	Answer to 4 sig. figs.
a	4124.7841	4125
b	38.4163	
c	291.7412	
d	39 042.611	
e	39 048.619	
f	38.4131	
g	446.982	
h	0.142937	
i	1317.699	
j	3050.1491	

There are many practical situations where careful measurement is essential – for example, dress-making, carpentry, weighing out parcels to calculate the cost of postage, and so on. However, in other situations, an *estimate* based on experience and common sense is often good enough. For example, when returfing a lawn, you may wish to measure its area fairly accurately using a tape measure, but if you decide to seed it, simply pacing it out to estimate the area may be sufficient. Estimation is a skill which greatly improves with practice. I sometimes find it helpful to imagine everyday objects of a standard size to help me make an estimate. For example:

Estimate	Helpful image
estimating height or distance	a door is roughly 2 metres high a running track is 400 metres around
estimating capacity/ volume	a standard milk bottle holds one pint
estimating weight	a bag of sugar weighs 1 kg
estimating air temperature	typical winter temperatures 0°C–10°C summer temperatures 20°C–30°C spring/autumn temperatures 10°C–20°C

And now, as promised earlier in the chapter, we turn to the units that are used in measuring.

Spotlight

The main advantage of metric units of measure is that they are standardized – so whether you live in Crewe, Colorado or Cairo, a metre is a metre is a metre. The classical definition of a metre is that it is one ten-millionth of the length of the Earth's meridian along a quadrant (i.e. the distance from the Equator to the North Pole). Since 1983, a more precise definition has been used: one metre is the distance travelled by light in free space in $\frac{1}{299\,792\,458}$ of a second. Now, if you'll excuse me, I'll just get out my school ruler and check that last digit!

Imperial and metric units

Until about 1970, measurement in the UK was largely done with imperial units. Since then, however, the British population has at last owned up to the fact that they have the same number of fingers and thumbs as the rest of the world, and have 'gone decimal'. The *decimalization* of money in 1971 was carried out quickly and effectively. As a result, most people mastered the new coinage within days. It was also intended that the familiar imperial units of length, weight and capacity be phased out within a few years. This change, called *metrication*, was to have swept away the most familiar of the measuring units – feet, inches, yards, pounds, stones, pints, gallons, and so on – in favour of metres, kilograms, litres and the like. Indeed, during the 1970s, many children learnt only the metric units in school on the assumption that the old imperial units would soon be six feet (sorry, 1.83 metres) under. Unfortunately, however, the changeover was so half-hearted that, at the present time of writing, we are still regularly using both systems (and having fun trying to convert from one to the other!). Since the 1980s, children have been taught both systems in school.

Conversions between metric and imperial units tend to involve rather awkward numbers. For example, there are 'about' 39.370078 inches in one metre! Not surprisingly, a number of half-baked approximations have appeared, like the metric yard, the metric foot and even the metric brick. Have a look at the table below and you will see just how 'approximate' some of the approximations are.

Unit	'True' value
Metric yard = 39 in	1 m = 39.370078 in
Metric inch = 2 cm	1 in = 2.54 cm
Metric foot = 30 cm	1 ft = 30.48 cm
Metric mile = 1500 m	1 mile = 1609.344 m

I have summarized most of the metric and imperial units that you are likely to need later in the chapter. I will explain how to use them by focusing on the most basic measure of all – length.

LENGTH

▶ **Metric units**

millimetre $\xrightarrow{\text{10}}$ centimetre $\xrightarrow{\text{100}}$ metre $\xrightarrow{\text{1000}}$ kilometre
(mm) (cm) (m) (km)

▶ **Imperial units**

inch $\xrightarrow{\text{12}}$ foot $\xrightarrow{\text{3}}$ yard $\xrightarrow{\text{1760}}$ mile
(in) (ft) (yd)

The numbers above the arrows tell you how to convert from one unit to another. Thus, there are 10 mm in 1 cm, 100 cm in 1 m, and so on. If you want to know how many mm are in 1 m, then *multiply* the two numbers 10 and 100 (i.e. there are 1000 mm in 1 m). It will help you understand and remember the metric units when you realize that:

▶ for each dimension there is a basic unit – the basic unit for length is the metre

▶ all the other units get their names from the basic unit, e.g. because centi- means one hundredth $(\frac{1}{100})$, then a *centimetre* is one hundredth of a metre.

The following table of metric prefixes will help you to work out the others.

Prefix	Meaning
Milli-	one thousandth $(\frac{1}{1000})$
Centi-	one hundredth $(\frac{1}{100})$
Deci-	one tenth $(\frac{1}{10})$
Kilo-	one thousand (1000)

Converting *between* metric and imperial units is a little trickier. If you don't need to be too accurate, it is helpful to remember that a 12-inch ruler is almost exactly 30 cm long. Dividing 30 by 12, it follows that one inch is roughly equal to 2.5 cm. When you need to be more accurate, use the conversion 1 inch = 2.54 cm, and also use a calculator!

AREA

Length is a *one-dimensional* (1-D) measurement because it involves only one direction. Problems involving surfaces (sizes of paper, carpets, curtain material, lawns …) are *two-dimensional* (2-D). Sometimes we describe area simply by stating the length and the breadth. For example, curtain material is bought by the metre (length) but we also need to know that the roll is 1 m 20 cm wide. If you are buying paint, on the other hand, the instructions on the tin may say something like: 'contents sufficient to cover 35 m²'. The unit described as a 'm²', or a 'square metre', is the basic metric unit of area. It means exactly what it says. One m² is the area of a square, 1 m by 1 m.

The area of a 1 m × 1 m square is equal to 1 m².

There should be enough paint in the tin to cover 35 of such squares.

Similarly, 1 ft² (1 square foot) is the area of a square 1 ft by 1 ft, and so on for the various other units of area. However, it's all not quite as easy as it sounds. Have a

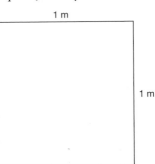

go at Exercise 7.5 now and see if you can avoid the traps that people often fall into.

Exercise 7.5 Area traps

a How many square feet (ft²) are there in one square yard (yd²)?

b How many cm² are there in one m²?

c What is the area of this rectangle?

d If you double the dimensions of this rectangle (i.e. double the length *and* double the breadth), what do you do to the area?

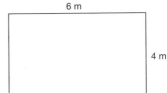

VOLUME

If you ask most people about the word 'volume', they will tell you that it is the knob on the TV set which makes it go loud and quiet. 'Volume', as used in mathematics, is rather different. It describes an amount of space in *three dimensions* (3-D). If you think of a box, its volume will depend on the three dimensions: length, breadth and height.

Unlike the words 'length' and 'area', 'volume' is not a word in very common everyday usage. We tend, instead, to use terms like:

> *How big* is the brick?

> What is the *size* of the box?

However, the trouble with words like 'big' and 'size' is that they don't necessarily refer to volume. In fact they can be called upon to describe any of a number of dimensions. For example, the size of a pencil might mean its length. The size of a piece of paper might mean its area. The size of a bag of sugar might even mean its weight, and so on.

People who work in the building trade become skilled at estimating amounts of earth and concrete. They usually measure these volumes in so many 'cubes'. A 'cube' usually refers to a cubic metre or a cubic foot. A cubic metre is the amount of space taken up by a cube measuring 1 m by 1 m by 1 m.

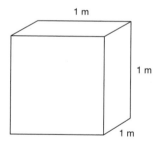

Similarly, a cubic centimetre (cc or cm^3) is the amount of space taken up by a 1 cm cube.

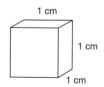

Another measure which deals with three dimensions is capacity. The difference between capacity and volume is that capacity describes a container, and is a measure to show how much the vessel holds. For example, we talk about the capacity of a saucepan, bucket, bottle, etc. The units of capacity, which are included in the tables below and over the following two pages, are normally only used with liquids.

Have a look now at the tables that follow, which show some of the most common metric and imperial units for length, area, volume, capacity and weight.

Spotlight

My cousin recently purchased his dream car – a BMW. His only disappointment was its fuel consumption, as displayed on the screen, which showed 31 mpg (for motorway driving, travelling at 60–70 mph). Not very impressive. It turned out that the on-board 'i-Drive' computer was set to record fuel consumption in miles per US gallon. Switching to the UK setting immediately improved this figure to a more respectable 37 mpg (note that 1 US gallon = 0.833 imperial gallons, approximately).

CONVERTING BETWEEN UNITS

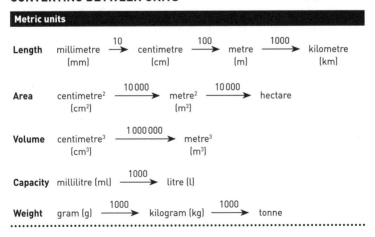

Metric units

Length   millimetre $\xrightarrow{10}$ centimetre $\xrightarrow{100}$ metre $\xrightarrow{1000}$ kilometre
(mm)   (cm)   (m)   (km)

Area   centimetre² $\xrightarrow{10\,000}$ metre² $\xrightarrow{10\,000}$ hectare
(cm²)   (m²)

Volume   centimetre³ $\xrightarrow{1\,000\,000}$ metre³
(cm³)   (m³)

Capacity   millilitre (ml) $\xrightarrow{1000}$ litre (l)

Weight   gram (g) $\xrightarrow{1000}$ kilogram (kg) $\xrightarrow{1000}$ tonne

Length inch (in) $\xrightarrow{12}$ foot (ft) $\xrightarrow{3}$ yard (yd) $\xrightarrow{1760}$ mile

Area inch² $\xrightarrow{144}$ foot² $\xrightarrow{9}$ yard² $\xrightarrow{4840}$ acre $\xrightarrow{640}$ square mile
(in²) (ft²) (yd²)

Volume in³ $\xrightarrow{1728}$ ft³ $\xrightarrow{27}$ yd³

Capacity fluid ounce $\xrightarrow{20}$ pint $\xrightarrow{2}$ quart $\xrightarrow{4}$ gallon

Weight ounce $\xrightarrow{16}$ pound $\xrightarrow{14}$ stone $\xrightarrow{8}$ hundredweight $\xrightarrow{20}$ ton
(oz) (lb) (cwt)

You will probably need practice at using these tables, so do
Exercise 7.6 now.

Exercise 7.6 Getting familiar with the units of measure

a How many millimetres are there in a metre?

b How many centimetres are there in a kilometre?

c How many grams are there in a tonne?

d How many inches are there in a mile?

e How many square inches are there in a square yard?

f How many ounces are there in a ton?

Unfortunately it isn't enough to be able to convert units within
the metric or the imperial system separately. Sometimes it is
necessary to convert between the two. The table below shows
some of the most common conversions between metric and
imperial units.

(*Note*: km/h means kilometres per hour.)

	Accurate conversion	Rough 'n' ready conversion
Length	1 m = 39.37 in	1 metre is just over a yard (a very long stride)
	1 in = 2.54 cm	1 inch = $2\frac{1}{2}$ cm
Capacity	1 litre = 1.76 pints	1 litre = $1\frac{3}{4}$ pints (a large bottle of orange squash)
	1 gallon = 4.54 litres	1 gallon = $4\frac{1}{2}$ litres
Weight	1 kg = 2.2 pounds	1 kg = just over 2 lb (a bag of sugar)
	1 pound = 0.454 kg = 454 g	1 lb = just under $\frac{1}{2}$ kilogram
Speed	100 km/h = 62.1 mph	8 km/h = 5 mph
	100 mph = 161 km/h	

Again, it would be a good idea to practise some of these conversions now, so have a go at Exercise 7.7.

Exercise 7.7 Practice exercise

Use the tables above and on the previous two pages (and a calculator where appropriate) to answer the following:

a Which of these would be a reasonable weight for an adult?
60 kg, 600 kg, 6 kg

b Is it better value to buy a 25 kg bag of potatoes or a 56 lb bag for the same money?

c What is the height of your kitchen ceiling from the floor, in metres?

d Some French roads have 90 km/h speed limits. What is this roughly in mph?

e Is $\frac{1}{2}$ litre of beer more or less than a pint?

f If we bought milk by the $\frac{1}{2}$ litre, how many bottles would you have to buy to have roughly 7 pints?

As a footnote, I feel that I should raise the issue of my use of the word 'weight' throughout this chapter. Strictly speaking, I should talk about 'mass' rather than weight. The weight of an object is a measure of the force of gravity acting on it, and this will vary depending on where the object is in relation to the Earth. Mass is the 'amount of matter' which the object contains and, wherever its position, this will not vary. Most scientists feel that this distinction is critical, but, provided your mathematics is conducted mostly on the Earth's surface, I wouldn't let it worry you too much!

Answers to exercises for Chapter 7

EXERCISE 7.1

Question	Dimension of measure	Likely units of measure
How heavy is your laundry?	weight	kg or lb
How long is the curtain rail?	length	cm or in
How hot is the oven?	temperature	degrees (°C or °F)
How far is it to London?	length/distance	km or miles
How fast can you run?	speed	km/h or mph
How long does it take to cook?	time	minutes
How much does the jug hold?	capacity	ml, pint or fl oz
How big is your kitchen?	volume	m^3 or ft^3
How big is the field?	area	hectare or acre

EXERCISE 7.2

Everyday activities	Measuring dimensions
Baking a cake	W; T; T°; C
Buying and laying a carpet	L; A
Checking the children's shoes	L
Setting out on a journey in good time	L; T; S

EXERCISE 7.3

stop, amble, walk, jog, sprint

lie, roll over, sit up, stand, walk

rayon, silk, wool, cotton, linen

EXERCISE 7.4

	Number	Answer to 4 sig. figs.
a	4124.7841	4125
b	38.4163	38.42
c	291.7412	291.7
d	39 042.611	39 040

	Number	Answer to 4 sig. figs.
e	39 048.619	39 050
f	38.4131	38.41
g	446.982	447.0
h	0.142937	0.1429
i	1317.699	1318
j	3050.1491	3050

EXERCISE 7.5

a $9 \text{ ft}^2 = 1 \text{ yd}^2$

b $10\,000 \text{ cm}^2 = 1 \text{ m}^2$

c Area = $6 \text{ m} \times 4 \text{ m} = 24 \text{ m}^2$

d Doubling the length and the breadth makes the area four times as big.

EXERCISE 7.6

a Number of millimetres in a metre = $10 \times 100 = 1000$

b Number of centimetres in a kilometre = $100 \times 1000 = 100\,000$

c Number of grams in a tonne = $1000 \times 1000 = 1\,000\,000$

d Number of inches in a mile = $12 \times 3 \times 1760 = 63\,360$

e Number of square inches in a square yard = $144 \times 9 = 1296$

f Number of ounces in a ton = $16 \times 14 \times 8 \times 20 = 35\,840$.

EXERCISE 7.7

a 60 kg would be a reasonable weight for an adult.

b A 25 kg bag weighs $25 \times 2.2 = 55$ lb. So, buying a 56 lb bag for the same money is a slightly better deal.

c A typical height of a kitchen ceiling from the floor is roughly $2\frac{1}{2}$ to 3 m.

d $90 \times 5 \div 8 = 56$ mph, roughly.

e 1 litre = 1.76 pints. So, since $\frac{1}{2}$ litre = $\frac{1.76}{2} = 0.88$ pints, this is less than one pint.

f 7 pints = $\frac{7}{1.76}$ = approximately 4 litres, or 8 half-litre bottles.

Summary

This chapter started by looking at the 'what', 'why' and 'how' questions of measurement.

What do we measure?   length, area, volume, capacity, weight, time, temperature, angle, speed ...

How do we measure?   – using words alone
                          – using words ranked in order (an ordering scale)
                          – using numbers (a number scale)

Why do we measure?   to help make decisions and comparisons

The final part of the chapter looked at the common metric and imperial units of measure, and at how we can convert within and between the two systems.

Finally, here is a checklist of the basic skills of measuring which you will need.

CHECKLIST

You should be able to:

▶ work with the common measures (length, weight, area, volume, capacity, time, speed, temperature)

▶ use and understand standard units of these measures (centimetre, kilogram ...)

▶ estimate lengths, weights and so on, in terms of these units, knowing when a measurement is about right and what sort of accuracy is appropriate

▶ use various measuring instruments (tape measure, ruler, weighing scales, balance, measuring jug, thermometer, clock)

▶ recognize composite units (miles per hour, price per gram, and so on).

To find out more via a series of videos, please download our free app, *Teach Yourself Library*, from the App Store or Google Play.

8

Statistical graphs

In this chapter you will learn:

▶ *how to draw a variety of statistical graphs and diagrams*

▶ *how to spot misleading graphs.*

Open a newspaper or watch the news on TV and you will be expected to make sense of a range of charts and graphs and to process statistical facts and figures. For example, here is a statistical fact.

> **Did you know that eight out of ten advertisers are prepared to mislead the public a little in order to sell their product?** Furthermore, the other two are prepared to mislead the public a lot!

One of the troubles with statistics is that there is such scope for deception. For example, I just made up the figures quoted above out of my head. But writing them 'in black and white' somehow seems to lend credibility to so-called 'facts'.

Deception can occur not only through the quoting of incorrect information. Equally common is 'dirty dealing' by means of the incorrect display of correct information. This chapter deals with the charts and displays that are most commonly used and misused in the media – barcharts, piecharts, line graphs and tables – as well as scattergraphs. It provides examples of where they are used, what they mean, how they are interpreted, and how they are sometimes misused to create a false impression.

Barcharts and piecharts

Barcharts and piecharts are useful when we want to compare different categories. Barcharts (sometimes called block graphs) consist of a set of bars set either vertically or horizontally. The height (or length) of each bar is an indication of its size.

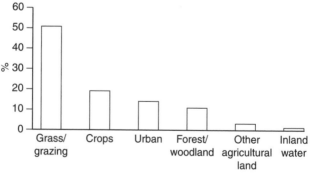

Figure 8.1 Land area by use.

Source: Social Trends 37, Figure 11.8, Office for National Statistics

Figure 8.2 Vertical barchart showing the lunchtime meals of pupils.

Source: Social Trends 25, Figure 3.13, CSO

The main strength of a barchart is that columns placed side by side, or one above the other, are easier to compare. So, in Figure 8.2 for example, you can see at a glance that the most popular form of lunchtime meal of pupils for the year in question was a packed lunch. You can also see that roughly twice as many children took a packed lunch than ate a free school meal.

Sometimes, barcharts can be used to show categories from more than one source on the same graph. The most common way of doing this is to use a compound barchart, as shown in Figure 8.3. This compound barchart shows two for the price of one – both male and female data are represented on the same graph.

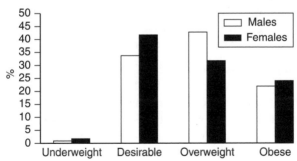

Figure 8.3 Compound barchart showing body mass by sex.

Source: Social Trends 37, Table 7.9, Office for National Statistics

Sometimes you may wish to draw a barchart where the category names are rather long. With a vertical barchart there simply isn't enough space to write the names beneath each bar, and so a horizontal barchart may be preferred.

Spotlight

Sometimes it is tempting to brighten up a chart or diagram with a few relevant images. However, this can sometimes produce a misleading chart. Look at this simple 'barchart' illustrating the results of a sports survey. 7 people were asked whether they preferred track or golf; 3 chose 'track' and 4 preferred 'golf'.

Can you see what went wrong here? It just happened that the icon for golf was much smaller than the one for track, as a result of which the column corresponding to 'track' is relatively too tall.

Track   Golf

Piecharts, as the name suggests, show the information in the form of a pie. The size of each slice of the pie indicates its value. For example, the two piecharts in Figure 8.4 depict the same data that were used for the compound barchart in Figure 8.3. In Figure 8.4, the data for males and females have been kept separate, with one piechart drawn for each. The piecharts show the number of males and females who fall into the four main categories of body mass (underweight, desirable, overweight and obese).

An important feature of a piechart is that it only makes sense if the various slices which make up the complete pie, when taken together, actually represent something sensible. For example, the piechart in Figure 8.5 shows, for different types of school, the various average pupil/teacher ratios; in other words the average number of pupils per teacher. As the figure title suggests, this is rather silly drawn as a piechart and would have been much better drawn as a barchart.

It is often too easy to cast your eye vaguely over a graph and murmur, 'Oh, yes, I see.' The next exercise asks you to linger on the various graphs that you have looked at so far and to make sure that you really do understand them.

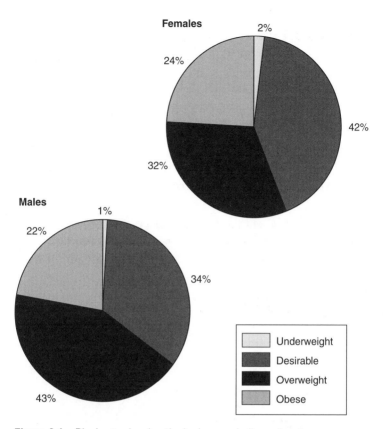

Figure 8.4 Piecharts showing the body mass indices of males and females.

Source: Social Trends 37, *Table 7.9, Office for National Statistics*

Type of school	Average ratio
Nursery	21:5
Primary	21:9
Secondary	15:4
Non-maintained	10:4
Special	5:8

Figure 8.5 A silly piechart showing pupil/teacher ratios by type of school, UK.

Source: Social Trends 25, *Figure 3.9, CSO*

Exercise 8.1 Interpreting graphs

1 In Figure 8.2, estimate the percentage of pupils falling into the four categories of the lunchtime meals. Use your estimates to check that they cover all of the pupils in the survey.

2 From Figure 8.2, estimate the percentage of pupils who ate a school meal, whether paid or free.

3 From Figure 8.3, would you say that it was men or women who had a greater than desirable weight?

4 From Figure 8.1, roughly what percentage of land is in non-urban use?

5 From Figure 8.4, which category of weight (underweight, desirable, overweight or obese) contains roughly a quarter of the females surveyed? For which category of weight are there roughly twice as many females as males?

6 Explain briefly in your own words why the piechart in Figure 8.5 is silly, and make a rough sketch of what the information would look like redrawn as a barchart.

Spotlight

In this era of decent computer applications, such as spreadsheets or graphing packages, the days of having to hand-draw a statistical chart are thankfully over. However, take care: just because something is easy to do on a computer doesn't mean that it is easy to do correctly. Remember that the computer is only as smart as the person driving it and the scope for using it to create inappropriate, incorrect or just plain silly charts is now vast!

Scattergraphs and line graphs

The graphs you have looked at so far have been helpful if you want to make comparisons – barcharts allow you to make comparisons based on the heights/lengths of the bars, while comparisons within piecharts are based on the relative sizes of the slices of the pie.

Sometimes, however, we wish to know how two different measures are related to each other. For example:

▶ Does a person's blood pressure relate to the fat intake in their diet?

▶ Is a child's health linked to the size of the family's income?

▶ How has a particular plant grown over time?

▶ Are lung cancer and heart disease linked to smoking?

To answer these sorts of questions, which look at two different measures together, we need a two-dimensional graph. This usually takes the form of either a scattergraph or a line graph.

The figures in Table 8.1 show the air temperatures and sales (in £) of an ice cream seller during 12 consecutive days in the summer. The same information is also portrayed in the scattergraph shown in Figure 8.6. By analysing the scattergraph, it is possible to explore the relationship between air temperature and sales. In particular, an obvious question is, 'Do sales of ice cream tend to be higher on warm days and lower on cool days?'.

Table 8.1   Air temperature and sales of ice cream over 12 consecutive days

Day	Temperature °C	Ice Cream Sales (£)
1	15	240
2	17	490
3	13	214
4	16	285
5	20	481
6	23	675
7	22	559
8	24	670
9	18	408
10	16	350
11	24	507
12	17	380

Source: made-up data

In order to draw a scattergraph of this information, you must place one measure on the horizontal axis, place the other on the vertical axis and mark on each axis a suitable scale. As you can see from Figure 8.6, I have chosen to place 'Temperature °C' on the horizontal axis and 'Sales' on the vertical axis.

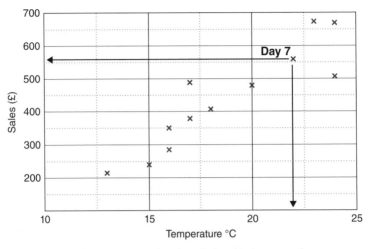

Figure 8.6 Scattergraph showing the relationship between air temperature and ice cream sales over 12 consecutive days.

The data for each day is plotted as a separate point. For example the point corresponding to Day 7 is highlighted on the graph. Following the down arrow to the Temperature axis, you can see that this point lines up with the value 22°C. Reading across to the Sales axis from the same point, the corresponding value is approximately £560 (the actual figure from the table is £559).

The pattern of points on the scatter graph helps to reveal the sort of relationship between the two measures in question. For example, in Figure 8.6 you can see that the points lie in a fairly clear pattern running from bottom left to top right on the graph. This reflects the not-to-surprising fact that ice cream sales tend to be high on warmer days and low on cool days. This is not a rigid relationship but more of a general trend. You will probably need to spend some time consolidating your understanding of a scattergraph, so do Exercise 8.2 now.

Exercise 8.2 Practising scattergraphs

Do you think that countries which have a high rate of marriage also tend to have a high divorce rate? Have a look at Table 8.2 and then at the corresponding scattergraph in Figure 8.7.

a Check that you understand how the points have been plotted and try to match each point up to its corresponding country.

b Which three countries show the lowest divorce rates? Can you provide a possible explanation for these low rates?

c Overall, does the scattergraph show a clear relationship between a country's marriage rate and divorce rate?

Table 8.2 Marriage and divorce rates per 1000 population: EU comparisons, 2002.

Country	Marriage	Divorces
Denmark	6.9	2.8
Netherlands	5.5	2.1
Portugal	5.4	2.6
Greece	5.2	1.1
Finland	5.2	2.6
Ireland	5.1	0.7
Spain	5.1	0.9
United Kingdom	4.8	2.7
France	4.7	1.9
Germany	4.7	2.4
Italy	4.7	0.7
Austria	4.5	2.4
Luxembourg	4.5	2.4
Sweden	4.3	2.4
Belgium	3.9	3.0
EU average	4.8	1.9

Source: Social Trends 34, Table 2.13, Office for National Statistics

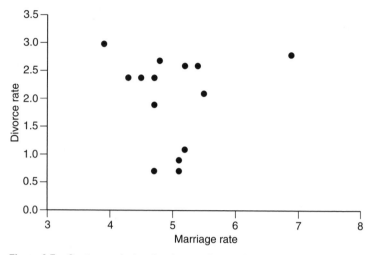

Figure 8.7 Scattergraph showing the marriage and divorce rate data.

Line graphs, like scattergraphs, are two-dimensional, so again we will be dealing with two measures at a time and examining the relationship between them. A line graph is one of the most common types of graph and indeed is what most people think of when we use the word 'graph'.

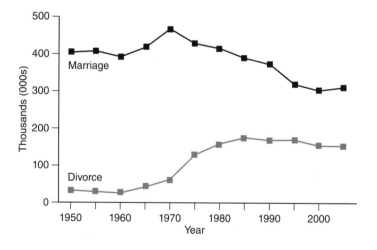

Figure 8.8 A time graph showing marriages and divorces in the UK.

Source: adapted from Social Trends 37, *Figure 2.9, Office for National Statistics*

Figure 8.8 shows a particular type of line graph, known as a *time graph*. It is so called for the obvious reason that the measure on the horizontal axis is time.

Exercise 8.3 Interpreting line graphs

a Estimate the number of marriages in the UK in 1992.

b Notice that the points marked on the two line graphs have been taken over five-year intervals. In the original line graphs (printed in the publication *Social Trends 37*), the points were based on data taken at one-year intervals. How do you think the size of the interval might affect the overall shapes of the line graphs?

c Overall, how have levels of marriage and divorce altered in the UK since 1950? How would you account for these patterns?

Misleading graphs

If you read through newspapers and magazines, it isn't difficult to spot graphs which are misleading. The graph below contains a fine collection of disasters! See how many you can spot.

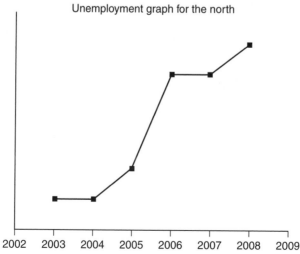

Unemployment graph for the north

Figure 8.9 Spot the errors on this graph!

In order to make sense of this graph and to see some of the distortions it contains, you really need to have a look at the data from which it was drawn (Table 8.3).

Table 8.3 Unemployment rates (%) for the north of England (2003 to 2008).

Year	Rate (%)
2003	4.8
2004	4.8
2005	4.9
2006	5.2
2007	5.2
2008	5.3

The graph shown in Figure 8.9 certainly looks dramatic. However, although unemployment rates in the north of England did rise over this period, the rise was not as dramatic as suggested by this representation. There are a number of errors and misleading features of the graph. Let's go through them in turn.

▶ The title is not very helpful. There is no clear explanation of what region the graph refers to ('the north' could refer to anywhere), or to what is being measured (the title should state that these are percentages).

▶ The axes are not labelled. The vertical axis should show clearly the figures and that they are 'Percentages', and the horizontal axis should say 'Year'.

▶ There is no scale marked on the vertical axis at all, so you have no idea what these figures are.

▶ Although there is no printed scale on the vertical axis, the scale fails to match up with the corresponding data in Table 8.3. In effect, the scale has been cut, which has the effect of making the graph look steeper.

A more correct version of the graph is shown in Figure 8.10.

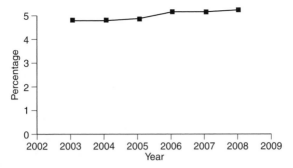

Figure 8.10 Line graph showing unemployment rates (%) for the north of England (2003 to 2008).

As you can see, now the increase is not nearly so dramatic and it is clear exactly what the figures refer to.

One final point about the scale on the vertical axis of Figure 8.9 is that it does not start at zero. In fact, it is acceptable to draw a vertical scale starting from a number other than zero, provided an indication is made on the axis that this has been done. The most common method is to mark a break on the axis, as shown below.

Figure 8.11 shows the same graph drawn with the vertical axis starting at 4.7% but with the break in the axis added to alert the reader to this potential source of confusion.

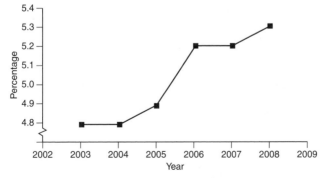

Figure 8.11 A graph demonstrating an axis break.

To end this section on misleading graphs, here is one of the most common types of distortion. Have a look at Figure 8.12 and see if you can spot how it might give a false impression.

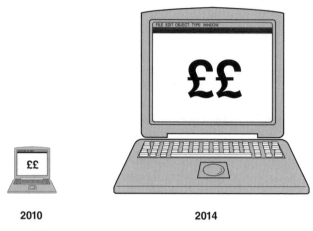

2010   **2014**

Internet sales to UK households in 2010 equals £27 billion
Internet sales to UK households in 2014 equals £107 billion

Figure 8.12   Internet sales to UK households.

Source: The *Guardian*, various articles

The diagram in Figure 8.12 shows up a favourite trick of advertisers, which is to make differences look bigger than they actually are. Certainly, internet sales increased greatly in the UK between 2010 and 2014 – by over four times in fact (the 2014 figure of £107 billion is roughly 4 times the 2010 figure of £27 billion). However, not only has the 2014 image been drawn four times as tall as the 2010 image, but it is also four times as wide. The overall impression of the larger image, therefore, is that it has an area which is 'four times four', i.e. sixteen times as great as that of the smaller image. Advertisers are able to exploit the fact that most of us work on impressions, not facts!

Over the next few weeks, why don't you look out for some more examples of misleading graphs in newspapers and magazines? Most people who present information to the public have some sort of vested interest. You will find it helpful to ask yourself: 'What *is* the vested interest, and therefore what impression is this graph designed to convey?'

If you are interested in finding out more about statistical ideas, why not get a copy of *Statistics: An Introduction*?

Answers to exercises for Chapter 8

EXERCISE 8.1

1 Estimates from Figure 8.1 are as follows:

Lunchtime meal	%
Packed lunch	32
Paid school meal	29
Free school meal	16
Other	23

2 The percentage of pupils who ate a school meal, whether paid or free, is 29% + 16% = 45%.

3 You must combine the 'Overweight' and 'Obese' categories to identify people with greater than desirable weight. This combined category produces a slightly higher percentage of males than females (65% compared with 56%).

4 Roughly 15% of land is in 'Urban' use, so the percentage in non-urban use is 85% (100% − 15%).

5 Roughly a quarter of the females (actually 24%) fell into the 'Obese' category. Roughly twice as many females as males fell into the 'Underweight' category.

6 The piechart in Figure 8.5 fails to meet a basic condition of a piechart in that the complete pie doesn't represent anything meaningful. In this case, the complete pie corresponds to the sum of the various average ratios in the different types of school, and who cares about that! The data would be more helpfully drawn as a barchart, as shown in Figure 8.13.

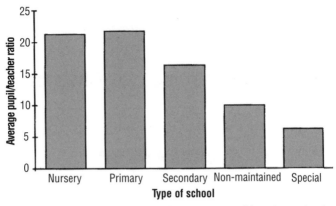

Figure 8.13 Vertical barchart showing average pupil/teacher ratios by type of school.

EXERCISE 8.2

a No comments.

b The three countries with the lowest divorce rates listed in Table 8.2 are Ireland, Spain and Italy. A possible explanation is that, at the time the data were collected (2002) these three countries were still strongly connected to their Catholic roots and so divorce was discouraged on religious grounds. However, these sorts of patterns can change quickly over time. To take an example, in Ireland in 2002, only 35 059 people were listed as divorced but just nine years later, in 2011, this number had risen to 80 770.

c There is little evidence of any clear pattern linking divorce and marriage rates in these points.

EXERCISE 8.3

a The number of marriages in the UK in 1992 was roughly 355 000.

b When graphs are plotted based on data taken at one-year intervals, as opposed to every five years, there are likely to be more subtle changes in direction. My graphs given in Figure 8.8 are actually rather crude, each being constructed by joining up 12 points with straight lines. However, as you will see in the answer to part c, the sudden blip in the divorce figures tends to be smoothed out when plotted over a five-year interval.

c   Marriage levels rose between 1950 and 1970 (as the post-war baby boomers reached marriageable age) and then fell (as cohabiting became a more acceptable alternative to marriage). Divorce remained fairly low until the early 1970s when it rose rapidly until the mid 1980s, when it levelled off and even dropped slightly. It was the 1971 Divorce Act that started this change (there were 80 000 divorces in 1971; by the following year the number had risen by over 50% to 125 000).

Summary

This chapter has covered four of the most common types of graphs: barcharts, piecharts, scattergraphs and line graphs. The final section dealt with misleading graphs, and a list was provided of some of the common ways in which graphs can be drawn in an unhelpful or deliberately distorted way.

CHECKLIST

Here are some important conventions to remember when drawing any graph or chart. Include:

▶ a helpful title

▶ labels and appropriate numerical scales for the axes and the units of measure

▶ data sources, where appropriate.

To find out more via a series of videos, please download our free app, *Teach Yourself Library*, from the App Store or Google Play.

9

Using a formula

In this chapter you will learn:

▶ *why algebra might be useful*

▶ *some of the rules of algebra*

▶ *how algebra can be used to prove things.*

Is algebra abstract and irrelevant?

For many students, the arrival of algebra in their school lessons was the point at which they felt they parted company with mathematics. Algebra has a reputation of being hard, largely because many people see it as abstract and irrelevant to their lives.

Let us first consider whether algebra is abstract. The simple answer to this charge is 'Yes it is!' Algebra is certainly abstract, for that is the point of algebra. The word 'abstract' means 'taken away from its familiar context'. The reason that algebra is such a powerful tool for solving problems is that it enables complex ideas to be reduced to just a few symbols. Naturally, if algebra is to be useful to you, you need to understand what the symbols mean and how they are related. Assuming this is the case, expressing something as a brief mathematical statement (which might be a formula or an equation) allows you to strip away the details, to forget about the context from which it was taken and focus on the essential underlying relationship. Of course, there are often situations where you *don't* want to strip away the context (in questions of human relationships, for example). Clearly, you wouldn't wish to use algebra for such problems.

Next, let's examine the charge of algebra being irrelevant. Most people believe that they never use algebra. Yet in many jobs, particularly, say, in medicine and engineering, formulas are crucial for converting units, calculating drug dosages, setting machines correctly for different tasks and so on.

Increasingly, large organizations and government institutions use formulas for deciding on and describing their funding arrangements. To take the example of education, a formula is used to define how much money is allocated to secondary and primary school budgets on the basis of the number and age of pupils. There are many questions that immediately arise. For example:

▶ Is it a fair way of allocating money?

▶ Is it right that the weighting (i.e. the relative amount) for secondary-age children is much greater than for primary-age children, or should all children be allocated the same amount, regardless of age?

▶ How can you find out what the relative weightings are for children of different ages?

▶ Who should decide these sorts of questions and can interested parents and teachers enter the debate?

The point I wish to make in introducing this chapter is that people can debate this question *only if they understand what a formula is saying*. If you don't understand basic algebra, other people will be making such decisions for you and you will have no idea whether or not they are acting in your best interests.

Before starting to examine any formulas, we begin by looking at some of the basic features of algebra – how it is used as a shorthand way of expressing something, and some of the conventions regarding how it is written.

Spotlight

The origins of algebra can be traced back four thousand years to the ancient Babylonians, who lived in what today we call Iraq. Despite the fact that they lacked any real algebraic notation, they were able to solve equations using methods very similar to those used today. For example, a maths textbook used in a Mesopotamian Elementary School would contain questions like:

'A quantity added to a quarter of itself is 20. What is the quantity?'

Algebra as shorthand

There are many situations in everyday life where it is convenient to adopt a shorthand – usually in the form of abbreviations or special symbols – in order to speed things up. I am aware that, for many people, the symbols in algebra seem to cause confusion rather than be an aid to efficiency. However, the idea of using a shorthand isn't just confined to mathematics.

Some examples for you to interpret are given in Exercise 9.1. Then, in Exercise 9.2 you are asked to examine some mathematical shorthands.

Exercise 9.1 Shorthands in everyday life

Example	Source	Meaning
a det. hse, lge gdns, gd decs, FCH ...		
b K1, P1, M1, C4F, K2 ...		
c PAS, MOT, fsh, good runner		
d 30s m, gsoh, wltm f for fta and poss ltr		

See if you can identify the source of these shorthands. What do they mean?

Exercise 9.2 Shorthands in mathematics

Here are some mathematical sentences. Rewrite each one in mathematical shorthand. The first one has been done for you.

Longhand	Shorthand
a Three multiplied by two and a half	$3 \times 2\frac{1}{2}$
b The sum of twelve, and four and three-quarters	
c The sum of the squares of three and four	
d Four times the difference of nine and three	
e Five plus four, all divided by the product of five and four	
f The number of inches, I, is found by multiplying the number of metres, M, by 39.37.	

As you can see from these examples, mathematics is full of shorthand notation. For example:

▶ Instead of writing numbers out as words, 'one', 'two', 'three', and so on, we have saved time by inventing the numerals 1, 2, 3, etc.

▶ Rather than say 'multiplied by' or 'added to', we use the symbols $\times$ and $+$.

▶ Squaring is represented by writing a small 2 above and to the right of the number or letter that is being squared. Thus, five squared, or 5 times 5, can be written as 5×5 or as 5^2.

More examples of notation will be explained in the next section. For now, let us focus on part **f** of Exercise 9.2, as it demonstrates a key feature of algebra.

The solution to this example, which is given with the answers for Chapter 9 (Exercise 9.2f), is repeated below for convenience.

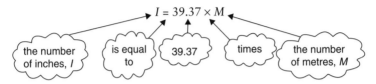

Before reading on, make sure that you can use this formula. For example, suppose you have just bought curtain material with a drop (the *drop* is typically the distance from the curtain rail to the window sill) of 1.20 metres and you want to know what that is in inches. Simply replace the *M* in the formula by 1.20 and calculate the corresponding value of *I*, as follows:

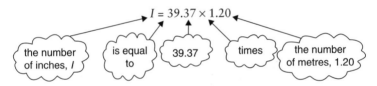

Using a calculator, the answer is 47.244 inches, a result which you might round up to 48 inches. (As an aside, rounding *up* may be appropriate here for two reasons. Firstly, if you are buying something like curtain material or a carpet, it is always better to have a little bit too much than to be a little bit short. Secondly, curtain material tends to come in various 'standard lengths' and, in imperial units, 48 inches happens to be one of these standard lengths.)

Now give some thought to how the formula, $I = 39.37 \times M$, has been written.

First, notice that I have introduced the abbreviations = and × to save the trouble of writing out the words 'equals' and 'times'. You may feel that this shorthand has reduced the formula to its bare essentials, but it is actually possible to dispense with the × altogether and write the formula even more briefly, as follows:

$$I = 39.37M$$

This demonstrates an important convention in algebra, namely that writing two letters together, or a number and a letter together, implies that they are multiplied. For example:

'*ab*' means '*a* times *b*'
'4*y*' means '4 times *y*'
'1.76*L*' means '1.76 times *L*'

and so on.

A second aspect of the formula worth noting is that I have used the letters *I* and *M* to represent, respectively, the number of inches and the number of metres. In algebra, the letters which we happen to choose to represent numbers are quite arbitrary. Thus, I could have written the formula as, say, $Y = 39.37X$, with Y representing the number of inches and X the number of metres. However, it is usually a good idea to relate each letter to the quantity that it represents, as this will help you to remember what the various letters stand for. For that reason, I used the initial letters of Inches and Metres, i.e. *I* and *M*, in this formula.

The most common letters used in basic school algebra tend to be

x, *y*, *a*, *b* and *n*

There is no obvious reason for this choice, with the possible exception of the *n*, which can be thought of as representing some unknown *n*umber.

Spotlight

I once observed a rather nice introductory lesson in algebra.

$$\square + 3 = 5$$

$$box + 3 = 5$$

$$\smile x + 3 = 5$$

$$x + 3 = 5$$

The next section deals with formulas in practical contexts and how to use them to do calculations.

Calculating with formulas

Drug dosages need to be carefully calculated and measured out. Giving too little of the drug means that the patient doesn't get the full benefit, but giving too much could be highly dangerous. The problem is greatly complicated when the drug is to be administered to a child, because clearly a dose that would be suitable for an adult would be too much for a young child. There needs to be a way of adjusting the dosages depending, perhaps, on the body weight or the age of the child concerned. One such formula, based on the child's age, is as follows:

$$C = D \times \frac{A}{A+12}$$

where C represents Child dosage, D represents adult Dosage and A is the Age of the child.

Written out in longhand, this formula means the following:

$$\text{Child dosage} = \text{Adult dosage} \times \frac{\text{Age}}{\text{Age} + 12}$$

Example 1

Let us take an example where the adult dose of cough medicine is 10 mg of linctus. What would be the appropriate dosage for a child aged six years?

Solution

First let us write down what we know.

$$D = 10, A = 6$$

With these numbers to hand, we are ready to calculate the child's dose, using the formula.

$$C = D \times \frac{A}{A + 12} = 10 \times \frac{6}{6 + 12} = 10 \times \frac{6}{18} = 3\frac{1}{3} \text{ mg}$$

Now here are some to try for yourself.

Exercise 9.3 Calculating dosages

Using the formula $C = D \times \frac{A}{A+12}$, calculate the dosages for the following situations.

a An adult prescription of a certain drug is 24 micrograms (µg). What would be an appropriate dose for a child aged ten years?

b An adult prescription of another drug is 200 µg. What would be an appropriate dose for a child aged four years?

We now move on to another formula, this time for converting temperatures. Temperatures in degrees Celsius can be converted to degrees Fahrenheit with the following formula.

$$F = 1.8C + 32$$

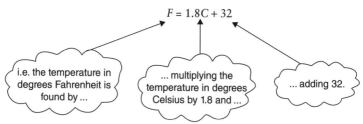

i.e. the temperature in degrees Fahrenheit is found by ...

... multiplying the temperature in degrees Celsius by 1.8 and ...

... adding 32.

Here is an example of the formula in operation.

Example 2 Oven temperatures

A typical cooking temperature for an oven is 180°C. What is this in °F?

Solution

Applying the formula:

The temperature in degrees Fahrenheit, $F = 1.8 \times 180 + 32$
$$= 356°F$$

Now here are some for you to try.

Exercise 9.4 Temperature conversion

a A warm summer day's temperature would be something like 30°C. What would this be in degrees Fahrenheit?

b The boiling point of water is 100°C. What is this in °F?

c There is only one temperature which is the same in degrees F as in degrees C. Try to find it.
(Hint: It is a temperature well below freezing point.)

The next example of a formula is concerned with calculating a phone bill. Telephone bills are usually calculated each quarter on the basis of a fixed sum for the rental of the line, plus a variable cost based on the calls you make. For example, my last bill of £97.61 was made up of a rental of £20.16 and a further 4.20 pence per unit used. The formula for this can be written as follows:

$$C = 20.16 + 0.042U$$

where C is the Charge in pounds and U is the number of Units used.

Notice that the charge rate of 4.20 pence per unit has been rewritten in the formula in pounds (i.e. as 0.042), in order to match with the units of the rental which is also expressed in pounds.

Example 3

I used 1844 units last quarter. Is the quarterly charge for my telephone bill correct?

Solution

The charge is calculated as follows:

$C = 20.16 + (0.042 \times 1844) = £97.61$ (rounded to the nearest penny)

This confirms the bill which I received as being correct.

Exercise 9.5 gives you the opportunity to try some of these for yourself.

Exercise 9.5 More bills

Calculate the quarterly charge for a household which used:

a 944 units

b 3122 units.

Spotlight

One formula that most people will have come across is Albert Einstein's equation connecting the energy, E, possessed by a body with mass m, and the speed of light, c. Einstein realized that matter and energy are really different forms of the same thing, so matter can be turned into energy, and vice versa. The formula $E = mc^2$ states that the energy possessed by a body is equal to its mass multiplied by the square of the speed of light.

Proving with algebra

This final section looks at an aspect of mathematics close to the hearts of mathematicians – the idea of proof. Without algebra, proving that a mathematical result is true is quite difficult. It is often easy enough to show that the result is true for several *particular* numbers, but it is quite a different matter to say that you *know* it is true for *all possible numbers*. For example, is it the case that adding two odd numbers always produces an even answer?

We could take some examples and see if it works. Thus:

3 + 7 = 10, which is even
5 + 11 = 16, which is even
23 + 15 = 38, which is even
111 + 333 = 444, which is even.

So, it does seem to be true, but have we proved it? Certainly not! Checking only four examples does not constitute a proof.

But what if I produced another 20 examples, or 100, or even a million? That would be quite impressive, but unfortunately providing lots and lots of special cases would not cut much mustard with a mathematician. Why, then, is it so hard to prove a numerical result to be always true? The reason is that you can't try all the infinite number of possible cases, and it would only take one of the ones you didn't get round to checking to be wrong to blow your theory apart. While arithmetic is useful for doing calculations with *particular* numbers, algebra is needed for making *generalizations*. The mathematical proof of this generalization (that the sum of two odd numbers is always even) is outlined and explained below. But before launching in to it, you need to spend a few minutes thinking about how we might represent even and odd numbers algebraically.

AN ASIDE ON EVEN AND ODD NUMBERS

If we think of a whole number as represented by, say, the letter K, then we can write even numbers as $2K$. You can check this out by giving K any value you wish to think of. For example:

▶ when $K = 3$, $2K = 6$, which is even

▶ when $K = 8$, $2K = 16$, which is even

▶ when $K = 13$, $2K = 26$, which is even

▶ when $K = 50$, $2K = 100$, which is even
   and so on.

The reason we know that $2K$ is always even is that it contains a factor 2, which is essentially what an even number is.

Similarly, if we represent a whole number by, say, the letter L, an odd number can be written with the formula $2L + 1$. Again, let's take a few examples to check this out:

▶ when $L = 3$, $2L + 1 = 7$, which is odd

▶ when $L = 8$, $2L + 1 = 17$, which is odd

▶ when $L = 13$, $2L + 1 = 27$, which is odd

▶ when $L = 50$, $2L + 1 = 101$, which is odd.

And again, from logical reasoning, we can show that the number $2L + 1$ must be odd. The explanation lies in the fact that the number $2L + 1$ is 1 more than the number $2L$, which itself must be even because it contains the factor 2. A number one greater than an even number is necessarily odd.

With these ways of representing even and odd numbers at our disposal, we are now ready to prove the earlier result algebraically. I have restated it in Example 4.

Example 4

Prove algebraically the result that the sum of two odd numbers always gives an even number.

Solution

One possibility might be to let both the two odd numbers be represented by $2K + 1$. However, the problem with doing this is that, whatever value for K is chosen, we find ourselves with two odd numbers with the same value. This is subtly different from the problem we set out to prove. We need to allow the two odd numbers to be different, so, using different letters, we can let them be $2K + 1$ and $2L + 1$, respectively.

Their sum is $(2K + 1) + (2L + 1)$.

Simplifying, we get $2K + 2L + 1 + 1 = 2K + 2L + 2$.

Now, notice that $2K + 2L + 2$ can be written as $2(K + L + 1)$.

Since this number contains a factor of 2, and $K + L + 1$ is a whole number, then $2(K + L + 1)$ must be even.

We have now proved the result *in general terms*. No matter what whole number values you think up for K and L, the general argument demonstrates that the result $2(K + L + 1)$ will always be an even number.

If you are unfamiliar with algebraic reasoning, you may need to read this proof through more than once. Then, when you are more confident, have a go at writing your own proof in your answer to Exercise 9.6.

By the way, don't worry if you find this difficult. Most people find algebraic proofs hard to fathom, and you would need a lot more practice at working with algebraic symbols than has been provided in this chapter if you are to perform proofs with confidence. I have included it mostly to indicate the sorts of things that mathematicians spend their time on, and to give you an insight into how algebraic symbols can be an aid in solving abstract problems.

Spotlight

Two words that cause confusion in algebra are *expression* and *equation*. Typically, an algebraic expression is just a collection of letters and numbers placed together with addition or subtraction signs between them. They are often provided in textbooks for students to practise collecting 'like' terms together ($2x$ and $3x$ are 'like' terms because the letter part, the x, is the same in each).

So, in the first expression below, the $2x$ and $3x$ terms can be put together giving $5x$ and the whole expression is simplified to $5x - 2y$.

Expressions

$2x - 2y + 3x$
$4a - 6b + 2b$

Equations consist of a statement of two expressions that are equal. You'll know it is an equation as it always contains an equal sign = (the clue is in the first four letters of '**equal**' and '**equa**tion').

Equations are provided in textbooks for students to solve – that is, to work out the value of the unknown letter that makes the equality true. For example, the solution to the first equation below is $x = 2$, because $2 + 3 = 5$.

Equations

$x + 3 = 5$

$2x - 5 = 3$

Exercise 9.6 Proving with algebra

Is it true that the product of two odd numbers is always an odd number? (*Reminder*: product means 'the result of multiplying together'.)

Test this out first with a few special cases and then try to prove it algebraically.

Answers to exercises for Chapter 9

EXERCISE 9.1

Example	Source	Meaning
a det. hse, lge gdns, gd decs, FCH …	Newspaper ad. for a house	Detached house, large gardens, good decorations, full central heating …
b K1, P1, M1, C4F, K2, …	Knitting	Knit 1, Purl 1, Make 1, Cable 4 forward, Knit 2
c PAS, MOT, fsh, good runner	Newspaper ad. for a car	Power assisted steering, holds an MOT certificate, full service history
d 30s m, gsoh, wltm f for fta and poss ltr.	Personal ad. in a US newspaper	Thirties male, good sense of humour, would like to meet female for fun, travel and adventure and possible long-term relationship.

EXERCISE 9.2

Longhand	Shorthand
a Three multiplied by two and a half	$3 \times 2\frac{1}{2}$
b The sum of twelve, and four and three-quarters	$12 + 4\frac{3}{4}$
c The sum of the squares of three and four	$3^2 + 4^2$
d Four times the difference of nine and three	$4(9-3)$
e Five plus four, all divided by the product of five and four	$\dfrac{5+4}{5 \times 4}$
f The number of inches, I, is found by multiplying the number of metres, M, by 39.37	$I = 39.37 \times M$

EXERCISE 9.3

a A ten-year-old's dosage is $24 \times \frac{10}{10+12} = 24 \times \frac{10}{22} = 10.9$ µg.

b A four-year-old's dosage is $200 \times \frac{4}{4+12} = 200 \times \frac{4}{16} = 50$ µg.

EXERCISE 9.4

a $F = 1.8 \times 30 + 32 = 86°F$

b $F = 1.8 \times 100 + 32 = 212°F$

c You might have tried to find this temperature by trial and error. The solution is -40. This can be checked by putting the value -40°C into the formula, and the result -40°F comes out.

Thus:
$F = 1.8 \times \text{-}40 + 32 = \text{-}40°F$

The answer can be calculated directly using algebra, as follows. Let the unknown temperature be T.

If the temperature $T°F = T°C$, then they are connected by the formula, as follows.

$$T = 1.8T + 32$$

The table below summarizes how this equation can now be solved.

Algebra	Explanation
$T = 1.8T + 32$	This is the equation to be solved.
$T - 1.8T = 1.8T - 1.8T + 32$	Subtract $1.8T$ from both sides (Note 1).
$-0.8T = 32$	Simplify the terms in T.
$\dfrac{-0.8T}{-0.8} = \dfrac{32}{-0.8}$	Divide both sides by -0.8 (Note 2).
$T = \dfrac{32}{-0.8}$	Simplify.
$= -40$	The solution of the equation is -40.

Note 1: The intention here is to collect the T terms on one side of the = and leave the number on the other side.

Note 2: The intention here is to isolate the T on its own. Remember that the object of the exercise is to find the value of T.

EXERCISE 9.5

The formula is $C = 20.16 + 0.042U$

a $C = 20.16 + (0.042 \times 944) = £59.81$ (rounded to the nearest penny).

b $C = 20.16 + (0.042 \times 3122) = £151.28$ (rounded to the nearest penny).

EXERCISE 9.6

It is true that the product of two odd numbers is always an odd number.

First, here are some special cases:

$3 \times 5 = 15$, which is odd
$5 \times 9 = 45$, which is odd
$13 \times 7 = 91$, which is odd
$113 \times 5613 = 634269$, which is odd.

Now we move on to a general algebraic solution.

As before, we can let these two odd numbers be represented by $2K + 1$ and $2L + 1$.

Their product is $(2K + 1) \times (2L + 1)$.

This can be written as $\quad 2K(2L + 1) + 1(2L + 1)$.

Simplifying, we get $\quad 4KL + 2K + 2L + 1$.

Ignoring the final term, the '1' for the moment, notice that the first three terms, $4KL + 2K + 2L$, can be written as $2(2KL + K + L)$.

Since this number contains a factor of 2, and $2KL + K + L$ is a whole number, then $4KL + 2K + 2L$ must be even.

Now add the final '1' and it follows that $4KL + 2K + 2L + 1$ is odd.

Summary

A key point made in the introduction to this chapter was that it is silly to criticize algebra because it is abstract. Essentially, the purpose of algebra *is* to be abstract. Algebra involves expressing relationships in a mathematical shorthand in the form of symbols and letters. This has the effect of reducing the problem to its bare essentials and allows you to see and manipulate its main features. Certain algebraic conventions were explained (for example, that writing $4X$ actually means '4 times X').

As the title suggests, the main activities of the chapter involved using formulas and you were invited to dip your toe into the esoteric world of mathematical proof.

To find out more via a series of videos, please download our free app, *Teach Yourself Library*, from the App Store or Google Play.

10

Puzzles, games and diversions

In this chapter you will learn:

▶ *about the fun side of maths with a collection of puzzles and number games.*

Spotlight

Curiosity may be a life-threatening trait for felines but it is the lifeblood of mathematics. There seems to be some characteristic of the human mind that makes it unable to resist the urge to tackle puzzles of every type. You only have to harness that problem-solving instinct, add a dash of confidence, and you will be on your way to becoming successful at mathematics. Indeed, it is surprising how often the act of thinking about how to solve a puzzle will lead you into thinking about some helpful piece of mathematics. And by the time you finish working through the puzzles in this chapter, you'll be heard to say more than once, 'That's another fine maths you've got me into!'

This chapter provides a few suggestions for number puzzles and activities, which should help to amuse and entertain on a long journey or a wet weekend.

1 Bus number game

Games with bus numbers are best played in cities, where buses are plentiful. See if you can spot bus numbers where:

▶ the digits add up to 10 (e.g. 163)

▶ all the digits are even (e.g. 284)

▶ all the digits are odd (e.g. 39)

▶ all the digits are prime numbers (e.g. 275).

With practice, most people quickly get good at spotting patterns in numbers, in which case the game can be made more challenging. For example, try to spot bus numbers that are the product of two primes (e.g. 91 is the product of 7 and 13, both of which are prime). Remember, product means multiply.

Spotlight: Interesting numbers

The renowned Indian mathematician, Srinivasa Ramanujan, was once visited by the British mathematician G.H. Hardy. Hardy remarked that he had just travelled in a taxi bearing the rather dull number 1729.

'On the contrary', said his friend, 'it is a very interesting number. It is the smallest number expressible as a sum of two cubes in two different ways' (10 cubed plus 9 cubed or 12 cubed plus 1 cubed).

Not everyone has quite the same fascination and skill with number properties as Ramanujan. But you should not underestimate the degree of interest and social cachet you are likely to attract at dinner parties, by passing on gems about the properties of certain numbers. For example, a perfect opening line during that awkward 'first introduced' phase at a party might be, 'Did you know that our host's telephone number is the first six digits of the decimal expansion of pi, backwards?'

Well, this sort of chat-up line certainly seems to work for me!

2 Pub cricket

This is played on a car or coach journey where the route is likely to pass a number of pubs.

Players take turns to 'bat'. You score according to the number of legs (human or animal) which can be seen on each pub sign that you drive past. For example, the 'Bull and Butcher' scores six (four for the bull and two for the butcher), while the 'White Hart' scores four. If you go past a pub sign which has no legs, then you are 'out' and the next player takes a turn at batting.

Popular signs in this game, by the way, are the 'Coach and Horses' (with 24 or more legs, depending on the number of horses) and the 'Cricketers' Arms' (with up to 30, depending on whether or not both batsmen and both umpires are depicted!)

3 Guess my number

One player picks a number between 1 and 100 and the other player must guess it with as few questions as possible. Note that the questions must be such that they require a Yes/No answer.

Useful questions are ones such as:
'Is it less than 50?' or 'Is it even?'
Less useful questions are ones such as:
'Is it 26?' since this eliminates only one number at a time.

4 Finger tables

Most people know their 'times table' up to about 5. However, with 6 and above they may have problems. Here is a method which provides the answer to all products between 6 and 10 (remember that in mathematics 'product' means what you get when you multiply), using the cheapest digital calculator around – your fingers.

Place your hands in front of you as in the diagram, thumbs uppermost.

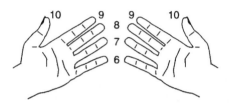

Number the fingers as shown above. Now to multiply, say, 7 and 8, touch the 7 finger of one hand with the 8 finger of the other (it doesn't matter which way round).

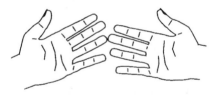

Now the answer to 8×7 can be found as follows:

a Count the number of fingers below and including the touching fingers (in this case 5). This gives the number of tens in the answer.

b Multiply the number of fingers on each hand above the touching fingers (here it is $3 \times 2 = 6$). This gives the number of units in the answer.

So the answer is 56.

Try it for some other numbers. Why does it always work? You will need to do some algebra to prove it works every time!

5 The story of 12

How many *different* stories can you make of 12?
Well, it is 11 + 1, 10 + 2, etc. Don't forget it is also 12 + 0.
What about 4 × 3, 3 × 4 and 6 × 2? And don't forget 12 × 1.
Now we're getting stuck. Ah! It's 24 lots of a half, so it's 24 × $\frac{1}{2}$.
Well, if you're allowing fractions it's $1\frac{3}{4}$ + $10\frac{1}{4}$.
Hey, this could go on all night!

6 Magic squares

This is a magic square.

8	1	6
3	5	7
4	9	2

A magic square is so named because all the rows, columns and diagonals 'magically' add to the same total (in this case, 15). This is known as a 3 by 3 magic square because there are three rows and three columns.

▶ Can you make a different 3 by 3 magic square so that all the rows, columns and diagonals add to 15?

▶ How about a 4 by 4 magic square … or a 5 by 5?

Hint: You may have spotted that the total of 15 for the 3 by 3 magic square is 3 times 5 (5 is both the number in the centre square and the middle value in the range of 1 to 9). Can you first of all work out what the rows, columns and diagonals of the 4 by 4 square should add to?

7 Magic triangle

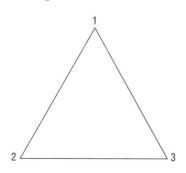

The numbers 1, 2 and 3 have been placed at the vertices (i.e. the corners) of the triangle.

Can you place two of the following numbers 4, 5, 6, 7, 8 and 9 along each side of the triangle so that the four numbers on each side (i.e. including the numbers at the vertices) add up to 17?

8 Upside down

The year 1961 reads the same when turned upside down.

a When was the most recent year prior to 1961 that reads the same upside down?

b When will be the next year that this works?

Explore what happens when letters and numbers are turned upside down. For example, what digits become letters of the alphabet when turned upside down?

9 Logically speaking

a What word is from this sentence?

b 1981 is to 1861 as 8901 is to what?

c Seven numbers can be seen in this sentence, but you know three of them are written out backwards. What are they?

d Ten birds sit on a roof. You make a noise to scare them off, and all but four of them fly away. How many are left?

e How old is a coin engraved with the date 88BC?

f You have three pairs of different-coloured socks in a drawer, each sock separated from its partner. How many single socks do you take out, without looking, to be sure that you have:

(i) a matching pair of any colour?

(ii) a matching pair of a particular colour?

g You need seven candle stubs to make a new candle. How many candles will you be able to make if you start with 49 stubs?

h What is the fewest number of coins you need to pay for something costing 85p without receiving change?

i Fill in the missing signs in the sums below.

2 ☐ 4 ☐ 3 = 5

2 ☐ 2 ☐ 2 = 3

8 ☐ 2 ☐ 2 = 2

10 The bells, the bells!

a If it takes 15 seconds for a church bell to chime 6 o'clock, how long does it take to chime midnight? (It's not as easy as it sounds!)

b A fence panel is 2 m long. How many fence posts are needed to panel a 16 m gap?

11 Calculator inversions

Enter the number 53045 on your calculator. Now turn it upside down. Can you see something that helps you keep your feet off the ground?

Next, try decoding the following 'message':

$3145 \times 10 + 123$

$204 \div 4$

$0.5 \times 0.5 \times 2$

257×3

Finally, have a go at the following crossword puzzle. Use the results of each calculation, upside down, to fit the puzzle.

Across

1 $3 \times 1000 + 45$ may be a good fit (4)

6 $5 \times 1111 - 18$ is no more (4)

7 $9^3 - 19$ is useful for troubled waters (3)

8 $2^5 + 11$ but please speak up! (2)

10 123×25 for a quick gin? (4)

12 Two score, for a surprise (2)

14 $13 \times 24439 + 5000000$, so stop and buy some (7)

16 $\frac{432}{6} + 1$ for the Spanish (2)

17 $\frac{70}{100}$ Verily, it sounds like a cow hath spoken (2)

Down

2 $0.65 + 0.1234$ is one third of a police officer's greeting (5)

3 $17 \times 10 \times 7 \times 3 + 3$ is another possibility (4)

4 $11^2 \times 31$ is emerald in Ireland (4)

5 $1111 \times 5 - 48$ for a no-win situation (4)

9 Two-fifths expressed as a decimal is one-third of a Christmas greeting (2)

11 1101×7 describes life in the 10 across lane? (4)

13 $19 \times 2 \times 193$ can be found on a 1 across (4)

15 Half could be a needle pulling thread (2)

12 Four 4s

The number 7 can be expressed using four 4s as follows:

$$7 = 4 + 4 - \frac{4}{4}$$

Express the numbers 0, 1, 2, ..., 10 using exactly four 4s and any other operation (for example, +, −, ×, ÷). You are also allowed square root, $\sqrt{}$.

Hint: 2 can be written as $\frac{4+4}{4}$.

13 Return journey

I plan to complete a return journey (there and back) in an average time of 40 mph. However, my outward journey is slow and I complete that part at 20 mph. How fast must I travel on the return journey to average 40 mph overall?

14 Find the numbers

a Two consecutive numbers add to give 49. What are the numbers?

b Three consecutive numbers have a total of 60. What are the numbers?

c Two consecutive numbers have a product of 600. What are the numbers? (*Note*: The product is what you get when you multiply.)

d Three consecutive numbers have a product of 1716. What are the numbers?

e Two numbers have a difference of 15 and a product of 54. What are the numbers?

15 Explore and explain the pattern

Try these out on your calculator.

a $37 \times 3, 37 \times 33, 37 \times 333$, etc.   (111, 1221, 12231)

b $1^2, 11^2, 111^2, 1111^2$ ....   (1, 121, 12321 ...)

c   Take three consecutive numbers.   (say, 8, 9 and 10)

Multiply the first and third number.   (8 × 10 =)

Square the middle number.   (9 × 9 =)

Subtract the smaller answer from the bigger answer.

The result is 1.   (81 − 80 = 1)

Does this always work for three consecutive numbers?

16 1089 and all that

Take a three-digit number.   (say, 724)

Reverse the digits and subtract whichever   (724 − 427 = 297)
is the smaller from the bigger.

Reverse the digits of the answer and add   (792 + 297 = 1089)
it to the answer.

Try it for other three-digit numbers. Why does 1089 keep cropping up? Do you always get 1089? If not, then which numbers does it not work for? Why?

Spotlight

By the way, this puzzle can be set up as a trick to impress your friends, as follows.

Write down the figure 1089 on a piece of paper and seal it in an envelope beforehand.

Then ask the friend to choose any three-digit number and perform the calculation described above.

Note: If a zero occurs in any part of the calculation, this must be counted as well. For example: 534 − 435 = 099.

Reversed, 099 becomes 990. This gives 099 + 990 = 1089.

In order to give the trick a little *pzazz*, ask your 'victim' for some additional but totally irrelevant information (for example, date of birth, telephone number, favourite colour, and so on).

17 Large and small sums

Try to arrange the digits 1, 2, 3, 4 and 5 (each used once only) to form two numbers so that the sum of the numbers is as large as possible.

Now try to arrange them so their sum is as small as possible.

Explore the same sorts of questions for multiplication, etc.

18 Gold pieces

In exchange for your lucky calculator and this oh-so-precious maths textbook, the evil, cunning and extremely wealthy empress Calcula offers you a choice of one of the following options.

You will be given:

either

a As many gold pieces as the number of minutes you have been alive;

or

b As many gold pieces as the largest number you can get on your calculator by pressing just five keys;

or

c One gold piece on the first day of this month, two on the second, four on the third, eight on the fourth, and so on, ending on the last day of the month.

With the help of your calculator, decide what you should do.

19 Initially speaking

This puzzle is easiest to explain with the following example.

Clue	Solution
3 B M, S H T R	3 Blind Mice, See How They Run

Now complete the solutions below.

If you get them all right, it should spell out the name of a film as well as a catchphrase when you read down the central boxes.

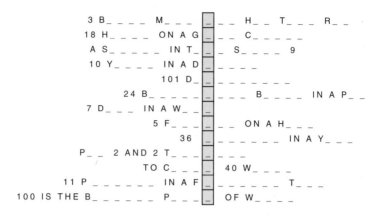

3 B_ _ _ _ M_ _ _ _ _ H_ _ T_ _ _ R_ _
18 H_ _ _ _ ON A G_ _ _ C_ _ _ _ _
A S_ _ _ _ _ IN T_ _ _ S_ _ _ _ 9
10 Y_ _ _ _ IN A D_ _ _ _ _
101 D_ _ _ _ _ _ _ _
24 B_ _ _ _ _ _ _ _ B_ _ _ _ IN A P_ _
7 D_ _ _ IN A W_ _ _
5 F_ _ _ _ _ ON A H_ _ _
36 _ _ _ _ _ _ IN A Y_ _ _
P_ _ 2 AND 2 T_ _ _ _ _ _
TO C_ _ _ _ 40 W_ _ _ _
11 P _ _ _ _ _ _ IN A F_ _ _ _ _ _ _ T_ _ _
100 IS THE B_ _ _ _ _ _ P_ _ _ _ OF W_ _ _ _

20 Guess the number

The rules of this game are given at the end of Chapter 5.

21 Nim

Nim is one of the oldest recorded games, possibly Chinese in origin, and is usually played by two people. There are many versions of Nim, one of which is described below.

Start with a pile of matchsticks. Each player, in turn, removes at least one but not more than six matches. The winner is the player who picks up the last match.

Here is a typical game.

Start with 28 matches in the pile.
A picks up 4 matches, leaving 24.
B picks up 5 matches, leaving 19.
A picks up 2 matches, leaving 17.
B picks up 6 matches, leaving 11.
A picks up 3 matches, leaving 8.
B picks up 1 match, leaving 7.

A (realizing that defeat is just a match away) picks up 1 match, leaving 6.

B picks up all remaining 6 matches, thereby winning game, set ... and match.

22 Calculator snooker

Player A enters any two-digit number. *Player B* takes a 'shot' by multiplying with another number. To 'pot' a ball, the first digit of the answer must be correct according to the table shown. (The degree of accuracy can be varied according to experience.)

Ball	Red	Yellow	Green	Brown	Blue	Pink	Black
Result needed	1...	2...	3...	4...	5...	6...	7...
Score	1	2	3	4	5	6	7

Otherwise, the rules are similar to 'real' snooker. There are 10 (or 15) reds and one of each of the six 'colours'. A player must score in the order red, colour, red, colour, and so on, until all the reds have gone. (*Note*: The colours are replaced but the reds are not.) When the last red has gone, the colours are potted 'in order' and are not replaced.

For example, one sequence of plays was:

Player	Enters	Display	Comments
Karen	69	69	
Peta	× 2 =	138	Peta pots the first red.
			She elects to go for blue ...
	× 5 =	690	... and misses.
Karen	× 2 =	1380	Karen pots the second red.
			She elects to go for black ...
	× 5.5 =	7590	... and pots it.
	× 1.6 =	12144	The third red.
			She elects to go for black again ...
	× 7 =	85008	... and misses.

23 Place Invaders

This game, for one or two players, can be played at different levels (1, 2, 3, etc.). Move on to a new level when you find the game too easy.

PLACE INVADERS 1

Enter a 3-digit number into the calculator. These three digits are removed one at a time by subtracting to zero.

Example: Starting number 352

Key presses	Display
− 2 =	350
− 50 =	300
− 300 =	0

You can make up your own numbers and let your partner remove them

Try the following:

416, 143, 385, 512, 853, 264, 179, 954, 589, 741.

PLACE INVADERS 2

This is the same as Place Invaders 1 except that the digits must be removed in ascending order.

Example: Starting number 352

Key presses	Display
− 2 =	350
− 300 =	50
− 50 =	0

i.e. you remove the 2, then the 3, and then the 5

PLACE INVADERS 3

This is the same as Place Invaders 2, except that you can use numbers with more digits. Try 4-digit numbers, then numbers with 5, 6, 7 and 8 digits.

PLACE INVADERS 4

This is the same as Place Invaders 3, except that you remove the digits by addition, not subtraction. This time the game will end with a 1 followed by a string of zeros. Use as few goes as possible.

Example: Starting number <u>1736</u>

Key presses	Display
+ 4 =	1740
+ 60 =	1800
+ 200 =	2000
+ 8000 =	10000

Note: If you start with a 5-digit number, the game ends with a display of 100000.

PLACE INVADERS 5

This is the same as Place Invaders 3, except that you can use decimals, e.g. 451.326, to be removed by subtraction in the order of 1, 2, 3, 4, 5, 6.

Note: If you make a mistake, it is easy to undo by adding back the number that you have just subtracted.

Answers for Chapter 10

Note: There are no comments for puzzles 1 – 5

6 MAGIC SQUARES

Here is another 3 by 3 magic square.

2	7	6
9	5	1
4	3	8

Any other solution must be a reflection or rotation of the original 3 × 3 square.

Now here is a 4 by 4 magic square.

16	3	2	13
5	10	11	8
9	6	7	12
4	15	14	1

With this 4 by 4 magic square each row, column and diagonal adds to 34. What makes this one even more magic is that each block of four corner squares also adds to 34.

Hang on – what about the four central numbers…?

7 MAGIC TRIANGLE

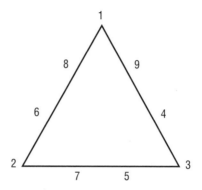

8 UPSIDE DOWN
a 1881   b 8008

9 LOGICALLY SPEAKING
a What word is **missing** from this sentence?

b 1981 is to 1861 as 8901 is to **1068.**

c The three 'backwards' numbers are in bold type here; the other four are underlined.
  <u>Seven</u> numbers can be se**en in** this sen<u>ten</u>ce, but you kno**w** <u>three</u> of them are writt**en** out backwards.

d Four birds are left.

e No coin could have been engraved with this date.

f (i) 4 socks   (ii) 6 socks

g Seven candles initially. But these will produce seven more stubs, so the answer is eight.

h Four coins: 50p + 20p + 10p + 5p

i $2 \boxed{\times} 4 \boxed{-} 3 = 5$

$2 \boxed{\div} 2 \boxed{+} 2 = 3$

$8 \boxed{\div} 2 \boxed{-} 2 = 2$ or $8 \boxed{\div} 2 \boxed{\div} 2 = 2$

10 THE BELLS, THE BELLS!

a Answer: 33 seconds.

It takes 15 seconds for 6 chimes. There are 5 intervals between the first and the sixth chime. Therefore it must take 3 seconds per interval. A series of 12 chimes has 11 intervals, hence 33 seconds.

b 9 posts are needed for 8 spaces.

Both the above questions refer to a common 'type' of maths question known as the old 'posts and spaces' trick. There are two things to remember in any 'posts and spaces' sort of situation. Firstly, be clear about which you are trying to count, the 'posts' or the 'spaces'. And secondly, remember that there is always one fewer 'space' than there are 'posts'.

11 CALCULATOR INVERSIONS

ELSIE IS SO ILL

	¹S	²H	O	³E			⁴I
⁵L		E		⁶L	E	S	S
⁷O	I	L		S			L
S		L		⁸E	⁹H		E
¹⁰S	¹¹L	O	E		¹²O	¹³H	
	O					E	
	¹⁴L	O	L	L	I	E	¹⁵S
¹⁶E	L					¹⁷L	O

12 FOUR 4S

$0 = 4 + 4 - 4 - 4$

$1 = \frac{4+4}{4+4}$ or $\frac{44}{44}$ or $\frac{4}{4} + 4 - 4$

$2 = \frac{4}{4} + \frac{4}{4}$

$3 = \frac{4+4+4}{4}$

$4 = 4 + \frac{4-4}{4}$

$5 = \sqrt{4} + \sqrt{4} + \sqrt{\frac{4}{4}}$

$6 = 4 + \frac{4+4}{4}$

$7 = \frac{44}{4} - 4$ or $4 + 4 - \frac{4}{4}$

$8 = 4 + 4 + 4 - 4$

$9 = 4 + 4 + \frac{4}{4}$

$10 = 4 + 4 + \frac{4}{\sqrt{4}}$

13 RETURN JOURNEY

It can't be done! Suppose the total distance (there and back) is 40 miles, then the total journey (there and back) must take exactly one hour. If the outward journey of 20 miles is completed at a speed of 20 mph, the one hour is completely used up!

14 FIND THE NUMBERS

a   24 and 25

b   19, 20 and 21

c   24 and 25

d   11, 12 and 13

e   3 and 18

15 EXPLORE AND EXPLAIN THE PATTERN

a   111, 1221, 12321. These numbers are palindromes (i.e. they read the same backwards as forwards).

b   1, 121, 12321, 1234321. Again a palindromic sequence similar, but not identical, to part **a**.

c   This result works for all sets of three consecutive numbers.

16 1089 AND ALL THAT

This result works for most but not all numbers. Try starting with some palindromic numbers and see what happens.

17 LARGE AND SMALL SUMS

The largest sum is 573 (541 + 32) or (542 + 31)

The smallest sum is 159 (125 + 34) or (124 + 35)

The largest product is 22 403 (521 × 43)

The smallest product is 3185 (245 × 13).

18 GOLD PIECES

Let us calculate each option in turn.

a As many gold pieces as the number of minutes you have been alive.
  Assuming that you are, say, 45 years old, the required calculation is:

  $45 \times 365 \times 24 \times 60 = 23\,652\,000$.

  In other words, between 23 million and 24 million.

b As many gold pieces as the largest number you can get on your calculator by pressing just five keys.
  My best effort here was to press:
  99 $\boxed{\times}$ 9 $\boxed{=}$
  This comes up with a puny 891 gold pieces. If your calculator has a 'square' function, marked $\boxed{x^2}$, you will do much better than this. So much so, that even five key presses will produce an answer too large for most calculators to display and which will therefore result in an error message. For example:
  99 $\boxed{x^2}$ $\boxed{x^2}$ $\boxed{x^2}$ produces an error message on my calculator.

c One gold piece on the first day of this month, two on the second, four on the third, eight on the fourth, and so on, ending on the last day of the month.
  This arrangement may sound very low key, but in fact it will produce an astronomically large result quite quickly. The best way to get an impression of the effects of doubling is to set your calculator's constant to multiply by 2, and then keep

pressing ☐=☐. What you will see should be something like the following:

Day	1	2	3	4	5	6	...	12	...	16
Amount	1	2	4	8	16	32	...	2048	...	32768

Before you get to the end of the month you will probably find that the calculator has over-stretched itself and produced an error message!

The answer, therefore, is that the 'best' option to choose depends on what sort of features your calculator has – for example, how many figures it displays, which keys it provides and so on. But whichever calculator you use, the third option is certainly a good one to go for!

19 INITIALLY SPEAKING

The phrase is SOME LIKE IT HOT.

```
       3 BLIND MICE   S  EE HOW THEY RUN
  18 HOLES ON A G     O  LF COURSE
      A STITCH IN TI  M  E SAVES 9
    10 YEARS IN A D   E  CADE
            101 DA    L  MATIANS
        24 BLACKB     I  RDS BAKED IN A PIE
    7 DAYS IN A WEE   K
          5 FING      E  RS ON A HAND
               36     I  NCHES IN A YARD
   PUT 2 AND 2 TOGE   T  HER
          TO CATC     H  40 WINKS
   11 PLAYERS IN A F  O  OTBALL TEAM
 100 IS THE BOILING POIN T  OF WATER
```

20–23 No comments.

To find out more via a series of videos, please download our free app, *Teach Yourself Library*, from the App Store or Google Play.

11

Spreadsheets

In this chapter you will learn:

▶ *what a spreadsheet is*
▶ *why spreadsheets are useful*
▶ *how to use a spreadsheet to calculate sums, sequences and percentages.*

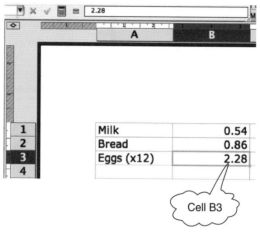

Figure 11.1   A basic spreadsheet grid

An overview of a spreadsheet

A spreadsheet is a computer tool that is used to set out information in rows and columns on a screen. Usually the rows are numbered 1, 2, 3, etc. down the left-hand side, while the columns are labelled A, B, C, etc. across the top. A typical spreadsheet might resemble that shown above, except that more rows and columns are visible on the screen at any one time.

Each 'cell' is a location where information can be stored. Cells are identified by their column and row position. For example, cell B3 is indicated in Figure 11.1: it is the cell in row 3 of column B.

The information you might want to put into each cell will be one of three basic types.

▶ Numbers: these can be either whole numbers or decimals, for example 7, 120, 6.32.

▶ Words: these can either be headings or explanatory text.

▶ Formulas: the real power of a spreadsheet is its ability to handle formulas.

Here is a simple example of using a formula.

First, I enter my height in metres (1.72) into cell A2: by selecting cell A2, typing 1.72 and pressing **Enter**. I could then enter a formula into cell B2 for converting the height into centimetres: by selecting cell B2 and typing:

=A2*100

Press the **Enter** key to 'enter' the formula and the value 172 is displayed in B2. Note that, although I have typed a formula into cell B2, what is displayed is the numerical value of the formula (in this case, 172).

This is illustrated in Figure 11.2 below. Note that on this screen the cursor is currently set on cell B2 (this is evident as both column B and row 2 are highlighted, as is the cell itself). The value displayed in the cell is 172, and the formula that produced it is shown in the formula bar at the top.

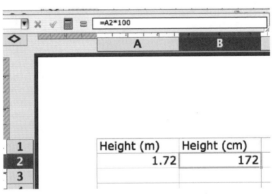

Figure 11.2 Converting metres to centimetres

In general, a formula entered into a particular cell will contain a calculation. More often than not, the formula will contain a reference to some other cell or cells. The cell containing the formula will then display a value calculated from the numbers currently stored in the cell (or cells) referred to in the formula. Now enter any new height in metres into cell A2 and the value in B2 immediately adjusts to display the corresponding height in centimetres.

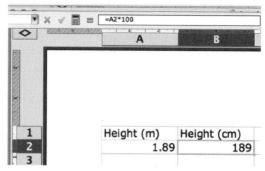

Figure 11.3 The calculation updates to match the new value in cell A2

A defining feature of a formula is that it starts with an '='
symbol. If you forget to include this, the spreadsheet will treat
the entry simply as text and display exactly what was typed in.
This is what has happened in Figure 11.4.

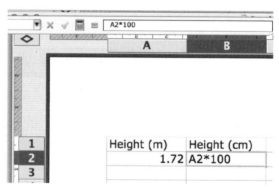

Figure 11.4 A formula with a missing '='

Spreadsheet formulas can cover anything from calculating
an average, or a row or column total, or perhaps, as in our
example, a conversion from one unit of measure to another.
As you can see, if relevant cell values are altered, the formula
will instantly and automatically recalculate on the basis of the
updated values and display the new result.

Spotlight

Every summer I teach at a maths summer school for teachers and one of the sessions is an introduction to spreadsheets. Many of the attending teachers have never used this tool before, and the standard comment afterwards is: 'I really had no idea you could do all those things on a spreadsheet!'

It is a common misconception that the primary function of a spreadsheet is for laying out data in a table so that it can be visually inspected. Yes, a spreadsheet will let you display data, but this is just a small part of what it can do for you. As well as being able to reorganize the figures, you can also use a spreadsheet to summarize them and plot them in a variety of different charts and graphs.

But most of all, as I hope you will find as you work through this chapter, using spreadsheets is fun!

Why bother using a spreadsheet?

A spreadsheet is useful for storing and processing data when repeated calculations of a similar nature are required. Next to word processing, a spreadsheet is the most frequently used tool in business. It is also extremely useful for householders to help solve problems that crop up in their various life roles as consumers, tax payers, members of community organizations, etc. A spreadsheet can be used to investigate questions such as:

▶ How much will this journey cost for different groups of people?

▶ Is my bank statement correct?

▶ Which of these buys offers the best value for money?

▶ What is the calorie count of these various meals?

▶ What would these values look like sorted in order from smallest to biggest?

▶ How can I quickly express all these figures as percentages?

A spreadsheet is a powerful tool for carrying out repeated calculations. It works on the following principle – you simply perform the first calculation, and then a further command will 'fill in' all the other calculations automatically. This applies whether you want to fill down a column or fill across a row.

Another advantage of a spreadsheet over pencil and paper is its size. The grid that appears on the screen is actually only a window on a much larger grid. In fact, most spreadsheets have hundreds of rows and columns, should you need to use them. Movement around the spreadsheet is also straightforward – you can easily move to adjacent cells or to any other cell of your choice.

Using a spreadsheet

Although users are not always aware of it, many home computers already have a spreadsheet package installed (it may be part of a larger suite of office applications). If you have access to a computer with a spreadsheet and want to make a start at using it, then read on. In this section, you will be guided through some simple spreadsheet activities. By the way, in order to illustrate spreadsheet principles, the examples presented here involve simple calculations using only three or four numbers. Please bear in mind that these are merely illustrative – the real power of a spreadsheet is experienced when these techniques are applied, at a stroke, to rows or columns containing ten or fifty or a thousand items of data.

There are several spreadsheet packages on the market and fortunately their mode of operation has become increasingly similar in recent years. However, your particular spreadsheet application may not work exactly as described here, so you may need to be a little creative as you try the activities below.

A very useful feature of any spreadsheet is that it can add columns (or rows) of figures. This is done by entering a formula into an appropriate cell. As you have seen, formulas are created by an entry starting with '='.

Exercise 11.1 Shopping list

Open the spreadsheet package and, if there isn't already a blank sheet open, create one by clicking on File and selecting New.

Using the mouse, click on cell A1 and type in the word 'Milk'. The word will appear on the 'formula bar' near the top of the screen.

Press **Enter** (the key may be marked **Return** or have a bent arrow pointing left). This causes the word 'Milk' to be displayed in cell A1.

Using this method, enter the data shown in Figure 11.5 into your spreadsheet.

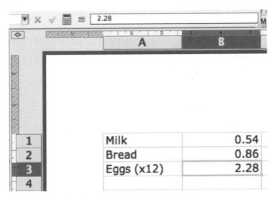

Figure 11.5   A shopping list with prices

Exercise 11.2 Finding totals using a formula

The formula for adding cell values together is:

   = SUM()

Inside the brackets you need to enter the cells, or cell range, whose values you want to add together.

Click in cell B4 and enter:

   = SUM(B1:B3)

The colon (:) here indicates a range of cells. So 'B1:B3' means cells B1, B2 and B3.

Spotlight

There are many useful shortcuts when entering a formula. Here is one I use when entering a formula that contains a cell range like this:

= SUM(B1:B3)

Enter the first part of the formula, = **SUM(**

Now, rather than typing the range, B1:B3, simply select the cells corresponding to this range of values. This will automatically enter B1:B3 into the formula. All that remains is to type the close bracket key and you're done!

By the way, it doesn't matter whether you type spreadsheet commands in upper or lower case. So you could have typed:

=sum(b1:b3)

Press **Enter** and the formula in cell B4 produces the sum of the values in cells B1 to B3. Next, enter the word TOTAL into cell A4.

Your spreadsheet should now look like the one shown in Figure 11.6.

	A	B
1	Milk	0.54
2	Bread	0.86
3	Eggs (x12)	2.28
4	TOTAL	3.68
5		

Figure 11.6 Using the =sum() command

Suppose that you gave the shopkeeper £10 to pay for these items. How much change would you expect to get? Again, this is something that the spreadsheet can find easily. Calculations involving adding, subtracting, multiplying and dividing require the use of the corresponding operation keys, respectively marked +, −, * and /. Note that the letter 'x' cannot be used for multiplication: you must use the asterisk*. You will find these keys, along with the number keys and '=', conveniently located on the numeric keypad, usually on the right-hand side of your keyboard (these keys are also available elsewhere on the keyboard, but you may have to hunt them down, which takes time).

Exercise 11.3 Calculating the change from £10

Enter the word 'TENDERED' in cell A5, the number 10 in cell B5 and the word 'CHANGE' in A6. Now enter a formula into B6 which calculates the change.

Remember that the formula must begin with an equals sign and it must calculate the difference between the values in B5 and B4.

Now try a slightly more complicated shopping list, this time with an extra column showing different quantities. You can enter this into the previous spreadsheet, a little lower down.

Enter the data in Figure 11.7 into your spreadsheet, starting with the first entry in cell A8.

	DESCRIPTION	QUANTITY	LIST PRICE	COST
8				
9	Pens black	24	0.37	
10	Folders	45	0.25	
11	Plastic tape	4	0.68	
12			TOTAL	

Figure 11.7 Two columns of figures

In order to calculate the overall total cost, you must first work out the total cost of each item. For example, the total cost of the black pens is $24 \times £0.37$.

Enter into cell D9 the formula:

= B9*C9

This gives a cost, for the pens, of £8.88. Now, you could repeat the same procedure separately for the folders and the plastic tape but there is an easier way, using a powerful spreadsheet feature called 'fill down'. You 'fill down' the formula currently in D9 so that it is copied into D10 and D11. The spreadsheet will automatically update the appropriate references for these new cells.

Click on cell D9 (which currently displays 8.88) and release the mouse button. Now move the cursor close to the bottom right hand corner of cell D9 and you will see the cursor change shape (it may change to a small black cross, for example). With the cursor displaying this new shape, click and drag the mouse to highlight cells D9:D11 and then release the mouse button. The correct costs for the folders (£11.25) and the plastic tape (£2.72) should now be displayed in cells D10 and D11 respectively. Now click on cell D10 and check on the formula bar at the top of the screen that the cell references are correct. Repeat the same procedure for cell D11. Magic!

Exercise 11.4 Summing up

With the costs in column D completed, you are now able to calculate the total cost. As before, this requires an appropriate entry in D12 using the = SUM() command. Do this now; you should get a total bill of £22.85.

Spotlight

Suppose you want to enter the sum of money £11.30 into the cell of a spreadsheet. When you do this, you'll find that the zero will be automatically removed and the value displayed as 11.3. This can be inconvenient. I find that, when displaying amounts of money on a spreadsheet, I prefer to fix the number of decimal places to two. This means that every value will displayed helpfully as pounds and pence, dollars and cents, and so on. It also means that all the numbers are decimally aligned, which makes them easy to compare as you run your eye down a column of figures.

To do this, first select the column or row of cells that you wish to format. Then find the command that lets you fix the number of decimal places (on most spreadsheets this facility will be a special button on the top of the screen). Instantly, all the numbers in the selected range will be displayed to two decimal places.

Number sequences on a spreadsheet

Having trouble remembering your 7 times table? Have no fear, because the spreadsheet can generate these sorts of sequences until the cows come home. We'll start by creating, in column A, the numbers 1, 2, 3, 4, ... and then, in column B, the corresponding numbers that form the 7 times table (7, 14, 21, 28, ...) Open a new worksheet (by clicking the 'Sheet2' tab at the bottom of the screen). In cell A1 enter the number 1. In cell A2, enter the following formula, which will display the value in A1, plus 1:

= A1+1

Select cell A2 once more and use the 'fill down' technique to fill this formula down as far as cell A30. You should now see the numbers 1 to 30 in this column.

Select cell B1 and enter the following formula, which will seed the 7 times table:

= A1*7

Reselect this cell and fill down as far as cell B30. Your spreadsheet should now look like the one in Figure 11.8.

In mathematics, multiplication is sometimes referred to as 'repeated addition'. For example, in the case of the 7 times table, this can be generated by repeatedly adding 7. As you will see from the next exercise, this suggests an alternative way of generating the 7 times table.

Exercise 11.5 Repeated addition

Enter the value 7 in cell C1.

Using a formula involving addition, generate in column C the numbers in the 7 times table (i.e. column C should display the same values as column B, but based on a formula using addition rather than multiplication).

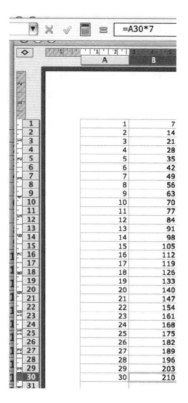

Figure 11.8 The 7 times table

Percentages on a spreadsheet

As you have already seen, a spreadsheet really comes into its own when you need to perform a large number of similar calculations. Essentially, you only need to set up the calculation once, applied to a particular cell value, and then use 'fill down' to apply the calculation to every value in the column. Here is a simple example involving calculating percentages. Return to the first spreadsheet that you created in Exercise 11.1. Here you listed the cost of milk, bread and eggs. Suppose that the shopkeeper decided to increase the price of each of these items by between 20p and 30p, as follows:

Item	Old price (£)	New price (£)
Milk	0.54	0.78
Bread	0.86	1.09
Eggs (×12)	2.28	2.53

When you are dealing with a higher-priced item such as a dozen eggs, a 25p price increase may not seem too much, but increasing the cost of a 54p bottle of milk by a similar amount is likely to raise more objection. The point is that, for lower priced goods, a 25p price increase represents a large percentage increase. Using a spreadsheet you can perform percentage calculations very easily and the grid format makes this information instantly understandable. The calculation is explained below in two stages: calculating the actual price increase for each item, and then working out what these increases are as a percentage of the old price.

CALCULATING THE ACTUAL PRICE INCREASE

Follow these steps now on your spreadsheet. Enter the new prices into column C as shown in Figure 11.9.

Select cell D1 and enter:

= C1–B1

Then press **Enter**. Select D1 again and fill down as far as cell D3. Column D now displays the actual price increases, in pounds (£).

	A	B	C	D
1	Milk	0.54	0.78	0.24
2	Bread	0.86	1.09	0.23
3	Eggs (x12)	2.28	2.53	0.25
4	TOTAL	3.68		
5	TENDERED	10		
6	CHANGE	6.32		

Figure 11.9 Finding the price rises

WORKING OUT THE PERCENTAGE PRICE INCREASES

Select cell E1 and calculate the percentage price increase for milk by entering:

$$= D1/B1*100$$

Select E1 again and fill down as far as cell E3. As shown in Figure 11.10, column E now displays the percentage price increases.

	A	B	C	D	E
1	Milk	0.54	0.78	0.24	44.4444444
2	Bread	0.86	1.09	0.23	26.744186
3	Eggs (x12)	2.28	2.53	0.25	10.9649123
4	TOTAL	3.68			
5	TENDERED	10			
6	CHANGE	6.32			

Figure 11.10   Finding the percentage increases

The percentage figures in column E are presented to many more decimal places than would be appropriate to this example. It is a simple spreadsheet task to round these numbers, so that they are displayed to, say, one decimal place or perhaps to the nearest whole number. Essentially, these price rises represented 44%, 27% and 11% of the original prices, respectively. This confirms what was stated earlier – when expressed in terms of percentage increases, a 24p price hike on a bottle of milk is much more significant than a similar price increase on something more expensive, such as a dozen eggs.

What else will a spreadsheet do?

This brief introduction has only really scratched the surface of what can be done with a spreadsheet. As was mentioned at the start of this chapter, the tiny data sets used here have been merely illustrative and don't properly reveal the power of a spreadsheet. Suppose, on the other hand, you have a table with 200 rows of data. Instead of having to perform 200 separate calculations, you only have to do one and you can 'fill down' the rest.

Once data have been entered into a spreadsheet, there are many other options available for summarizing and analysing some of the underlying patterns. For example, columns or rows can be reordered or sorted either alphabetically or according to size. Column and row totals can be inserted. As well as calculations using the four operations +, −, × and ÷, which can be found directly on the keyboard, a variety of other functions are also available within the spreadsheet's many menu options. These will enable you to calculate means, modes, medians and much more, at the touch of a button. The good news is that you won't have to remember the various commands needed to calculate these; they can be pasted directly from the appropriate menu option (possibly named 'Function' or something similar). An advantage of selecting functions via the menu is that they will be made available in a user-friendly way, so that the command syntax is made apparent.

Most spreadsheets have powerful graphing facilities which also allow you to select either all or some of the data and display them as a piechart, barchart, scattergraph, line graph, and so on. The detailed operation of the graphing facilities varies from one spreadsheet package to another, so they are not explained here. However, if by now you have greater confidence with using a spreadsheet, this is something that you might like to explore for yourself.

Finally, if you are interested in mastering the most popular spreadsheet application, Microsoft Excel (which is part of the Microsoft Office suite), get yourself a copy of *Teach Yourself Excel*.

Spotlight

If you don't have a spreadsheet application on your computer, don't despair. There are now several well-developed, web-based spreadsheets that are freely available – all you have to do is get onto the relevant website and sign up. A particularly good one is provided by Google; simply fire up any web browser, type into its search box the words 'Google spreadsheet' and then follow the relevant leads on offer.

Answers to exercises for Chapter 11

EXERCISE 11.3

The required formula for cell B6, which completes the table below, is:

= B5−B4

Your final spreadsheet table should look like this.

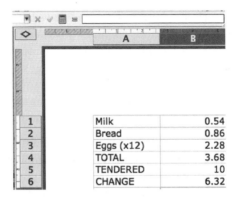

	A	B
1	Milk	0.54
2	Bread	0.86
3	Eggs (x12)	2.28
4	TOTAL	3.68
5	TENDERED	10
6	CHANGE	6.32

EXERCISE 11.4

You need to enter, in cell D12, the formula:

=SUM(D9:D11)

EXERCISE 11.5

You need to enter, in cell C2, the formula:

= C1+7

Then reselect cell C2 and fill down as far as cell C30.

Summary

▶ A spreadsheet is a computer tool that is used to set out information in rows and columns on a screen.

▶ Most home computers already have a spreadsheet package installed.

▶ They are useful for storing and processing data when repeated calculations of a similar nature are required; essentially, you only need to set up a calculation once.

▶ Once you have input all of your data into a spreadsheet you can summarize, analyse and display the results in a variety of ways.

To find out more via a series of videos, please download our free app, *Teach Yourself Library*, from the App Store or Google Play.

12

Diagnostic quiz

In this chapter you will learn:

▶ *how much maths you have already learned by reading this book!*

Now that you have carefully read through every page of the preceding 11 chapters (well, maybe you skipped a few pages!), you might like to take stock of what you have learnt, by trying to answer the questions in this diagnostic quiz.

A quiz, or test, can be tackled in many different ways. If you get every question right, you may feel good about yourself, but you probably haven't learned anything from it. If you get all the questions wrong, you will probably feel pretty depressed and unable to exploit the learning opportunities offered by the experience. I hope that you will be somewhere in-between. This quiz is not designed to trick you or to make you feel depressed. Having said that, you are very unlikely to find all the questions easy or to get every question right.

Here are some guidelines for tackling the quiz.

▶ You should be prepared to use your calculator for every question, except Question 1 where you are asked not to.

▶ Read each question carefully before you do it, so that you answer exactly the question that has been asked. For example, if it asks you to write numbers in order from smallest to largest, don't give your answers from largest to smallest.

▶ Don't be afraid to look things up in earlier chapters of the book if you have forgotten, say, how to convert miles into kilometres. This isn't a test to be taken under examination conditions, and you aren't expected to remember all the formulas and conversions in your head.

So, please give the quiz your very best shot. It is designed to take about one hour, but be prepared to take longer than that if you need to. When you have done all that you can do, then work through my solutions at the end of the chapter. As you will see, I have included detailed comments after the solutions, in order that you can 'turn your errors into learning opportunities'. For each one that you answered incorrectly, ask yourself the following questions:

▶ 'Where and why have I gone wrong?'

▶ 'What can I learn from this?'

Good luck!

Quiz

1 Try these calculations without using your calculator.

    a (i) 3×14 (ii) $-5 - 17$ (iii) $-20 \div 4$ (iv) $15 \div 75$

    b (i) $2\frac{1}{2} + 1\frac{1}{4}$ (ii) $3\frac{1}{2} - 2\frac{3}{4}$ (iii) $5\frac{1}{2} \times 3$ (iv) $3\frac{2}{3} \times 5$

    c (i) 8^2 (ii) $\sqrt{81}$ (iii) $\sqrt{(3^2 + 4^2)}$

2 Express the following as decimal numbers.

    a $20 + 7 + \frac{3}{10} + \frac{6}{100} + \frac{4}{1000}$

    b $60 - 3 + \frac{8}{10} + \frac{2}{100} - \frac{7}{1000}$

    c $\frac{62\,341}{1000}$

3 In the number 13.873, the '7' represents the number of hundredths. What does the '7' represent in the following numbers?

    a 271.93

    b 11.724

    c 0.00117

4 Which is bigger:

    a three-quarters or 70 per cent?

    b 0.06 or one-twentieth?

    c two-fifths or 0.5?

    d 10 per cent or an eighth?

    e 8 per cent or a tenth?

5 Using suitable metric units, estimate the following:

    a the height of a chair seat from the floor

    b the width of a cooker

    c the weight of a newborn baby

    d the distance from London to Birmingham

    e the capacity of a doorstep milk bottle

    f the weight of a letter

    g the temperature inside a domestic refrigerator

    h two teaspoonfuls of liquid

    i the thickness of a £1 coin

    j the temperature on a hot summer's day in London.

6 Place the following in order of size from smallest to largest:
   a 450 ml; half a litre; one pint; 75 centilitres
   b $1\frac{3}{4}$ metres; 18 cm; 1.2 km; 300 mm
   c half a week; four days; 95 hours; 0.01 of a year.

7 Using both imperial and metric units, estimate the following:
   a the speed of a car travelling on the outside lane of a
      motorway in the UK
   b the speed of someone having a brisk walk
   c the speed of a top 100 m runner
   d the speed of a supersonic jet aircraft.

8 A web-based store is holding a sale where each item is
   reduced by 30%. However, although VAT must be included
   in the payment, the prices they quote do not include VAT
   (at 20%). Postage and packing are free for payments over
   £30. How much would you expect to pay for an item that
   was originally quoted by the store (pre-sale) at £40?

9 The table below lists, in thousands of pounds (£000), the
   voluntary cash donations to the top 10 UK charities in a
   particular year.

Name	Income (£000)
National Trust	78 745
Oxfam	58 972
RNLI	56 229
Save the Children Fund	53 866
Imperial Cancer Research Fund	48 395
Cancer Research Campaign	45 352
Barnardo's	36 452
Help the Aged	33 141
Salvation Army	32 303
NSPCC	30 818

   a Rewrite the ten incomes, rounded to the nearest
      £million.
   b Using the rounded figures, sketch a horizontal barchart
      to represent the earnings of the top five charities in a
      particular year.

c Explain why a piechart would not be appropriate for depicting the earnings of the top five charities.

10 In a certain year, an estimated 19.2 million international visitors came to Britain and spent £9.2 billion. The piechart below shows the estimated annual tourist spending, broken down by which part of the world the tourists came from.

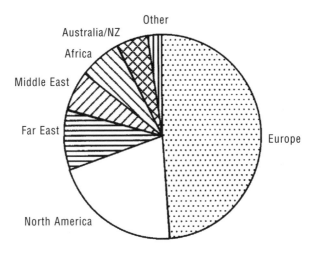

a From the piechart, estimate the annual spending by tourists from Europe.
b From which one of the regions listed here did roughly £2 billion of the tourist revenue come?
c If Britain were able to attract an extra 10 million visitors, how much more revenue might it be able to generate? Make a note of any assumptions that you have made in doing this calculation.

11 A survey of toy prices was taken in three large stores. The prices (in £) of three toys are summarized below.

	Hamleys	John Lewis	Toys Я Us
Toy A	12.99	9.25	9.29
Toy B	29.00	22.75	22.99
Toy C	7.99	5.95	4.97

On the basis of these prices:

a Which of the three stores seems to be the most expensive?

b Which store is the cheapest?

c If you bought all three items at the cheapest price on offer, how much would you have saved compared with buying them at the most expensive price?

d From your answer to part **c**, calculate your total savings as a percentage of the total cheapest price.

12 Petra earns £6200 per year doing part-time work. She pays tax at the basic rate of 20% and her tax allowances are £3750. The annual amount that she has to pay in Tax, T, can be calculated using the following formula.

$$T = 0.2\ (I - A)$$

(where I represents her annual income and A represents her tax allowances)

a Explain in your own words how, according to the tax formula, annual tax on earnings is calculated.

b Calculate how much Petra will have to pay in tax:
   (i) over the year
   (ii) each week. (Give your answer to the nearest penny.)

c Assuming there are no other stoppages from her wages, calculate how much Petra will receive each week after the weekly tax bill has been paid. (Give your answer to the nearest penny.)

Solutions to the quiz

1 **a** **(i)** 42   **(ii)** -22   **(iii)** -5   **(iv)** $\frac{1}{5}$ or 0.2

   b **(i)** $3\frac{3}{4}$ **(ii)** $\frac{3}{4}$ **(iii)** $16\frac{1}{2}$ **(iv)** $18\frac{1}{3}$

   c **(i)** 64   **(ii)** 9 (or -9)   **(iii)** 5 (or -5)

2 **a** 27.364
   b 57.813
   c 62.341

3 **a** tens
   b tenths
   c hundred-thousandths

4 **a** three-quarters
  b 0.06
  c 0.5
  d an eighth
  e a tenth

5 For this question, you must get *both* the correct answer *and* the correct units.
  a about 45 cm (I will accept answers between 40 and 50 cm)
  b about 60 cm, or 600 mm (I will accept answers between 50 and 70 cm)
  c about 4 kg, or 4000 g (I will accept answers between 3 and 5 kg)
  d about 175 km (I will accept answers between 150 and 200 km)
  e about 570 ml (I will accept answers between 500 and 600 ml)
  f about 40 g (I will accept answers between 10 and 60 g)
  g between 0°C and 5°C
  h about 12 ml (I will accept answers between 10 and 15 ml)
  i about 3 mm (I will accept answers between 2 and 4 mm)
  j about 30°C (I will accept answers between 25 and 35°C)

6 **a** 450 ml, half a litre, one pint, 75 centilitres
  b 18 cm, 300 mm, $1\frac{3}{4}$ m, 1.2 km
  c half a week, 0.01 of a year, 95 hours, four days.

7

		Imperial	*Metric*
a	the speed of a car on the outside lane of a motorway in the UK	70–90 mph	110–145 km/h
b	the speed of someone having a brisk walk	3–5 mph	5–8 km/h
c	the speed of a top 100 m runner	20–25 mph	30–40 km/h
d	the speed of a supersonic jet aircraft	above 760 mph	above 1200 km/h

8 Price including VAT = £40 × 1.20
Price after a 30% reduction = £40 × 1.20 × 0.7 = £33.60.
As payment exceeds £30, postage and packing are free, so
the total bill = £33.60

9 a The rounded incomes are as follows.

Name	Income (rounded to the nearest £million)
National Trust	79
Oxfam	59
RNLI	56
Save the Children Fund	54
Imperial Cancer Research Fund	48
Cancer Research Campaign	45
Barnardo's	36
Help the Aged	33
Salvation Army	32
NSPCC	31

b Below is a horizontal barchart showing the earnings of
the top five charities in the year in question.

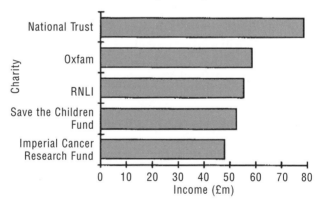

c A piechart would not be appropriate because the
combined income from these five charities together, which
would correspond to the complete pie, does not represent
anything meaningful.

10 a Roughly £4.5 billion (I will accept anything between
£4 billion and £5 billion.)

b North America

c Roughly £5 billion. This calculation assumes that the extra 10 million visitors spend at the same rate as do current visitors.

11 a Hamleys is the most expensive.

b John Lewis and Toys ЯUs are very similar in price, with Toys ЯUs having the slight edge (a total price of £37.25, compared with John Lewis' £37.95).

c Most expensive prices = £12.99 + £29.00 + £7.99 = £49.98
Least expensive prices = £9.25 + £22.75 + £4.97 = £36.97
Saving = £49.98 – £36.97 = £13.01

d Percentage saving = £13.01/£36.97 × 100 = 35.2%

12 a The annual tax bill, T, can be found as follows. Subtract the tax allowances, A, from annual income, I, and multiply the result by 0.2.

b (i) £490   (ii) £9.42

c £109.81

Detailed comments on the solutions

QUESTION 1

a (i) 3×14

This can be written out and calculated as follows:

```
    14
×    3
    42        Solution 42
```

(ii) Adding and subtracting with negative numbers can be confusing and it is sometimes a good idea to write the calculation out on a number line, as follows:

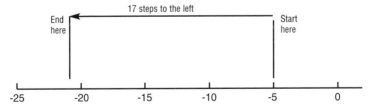

You start at -5 and then subtract 17. In other words, you move 17 steps to the left. This gives the result of -22.

(iii) $-20 \div 4$

Here you are dividing a negative number (-20) by a positive one (4). First you must decide on the sign of the answer (i.e. whether the answer is positive or negative). Because the two numbers are of different signs, the result must be negative, so write down the negative sign, -.

Next you do the calculation with the numbers.

$20 \div 4$ means $\frac{20}{4}$, giving the result 5.

So the solution is -5.

(iv) $15 \div 75$

This can be written as $\frac{15}{75}$.

The fraction can now be simplified to the simplest equivalent fraction by dividing the numerator and the denominator by 15, giving the solution $\frac{1}{5}$ or 0.2.

A common mistake here is to do the division the wrong way round, giving $\frac{75}{15} = 5$.

b **(i)** $2\frac{1}{2} + 1\frac{1}{4}$

First add the whole numbers: $2 + 1 = 3$.

Next, add the fraction parts: $\frac{1}{2} + \frac{1}{4}$. Remember that, in order to add fractions with different denominators, you must rewrite them as equivalent fractions which have the same denominator, which in this case is easiest done using quarters.

So, the fractions become $\frac{2}{4} + \frac{1}{4} = \frac{3}{4}$

Finally, add the fraction total to the whole number total.

Solution: $3 + \frac{3}{4} = 3\frac{3}{4}$

(ii) $3\frac{1}{2} - 2\frac{3}{4}$

First subtract the whole numbers: $3 - 2 = 1$.

Write down what still has to be calculated: $1\frac{1}{2} - \frac{3}{4}$

Notice that you can't just subtract the fraction parts directly because the fraction being subtracted (three-quarters) is bigger than the fraction it is being subtracted from (a half). The way around this is to borrow the whole number part, the 1, and turn it into quarters along with the fraction parts, as follows.

Solution: $1\frac{1}{2} - \frac{3}{4} = \frac{6}{4} - \frac{3}{4} = \frac{3}{4}$

(iii) $5\frac{1}{2} \times 3$

Multiply each part separately by 3 and then add the results together.

$5 \times 3 = 15$

$\frac{1}{2} \times 3 = 1\frac{1}{2}$

$15 + 1\frac{1}{2} = 16\frac{1}{2}$

(iv) $3\frac{2}{3} \times 5$

Multiply each part separately by 5 and then add the results together.

$3 \times 5 = 15$

$\frac{2}{3} \times 5 = \frac{10}{3} = 3\frac{1}{3}$

$15 + 3\frac{1}{3} = 18\frac{1}{3}$

c **(i)** 8^2, or 8 squared, means $8 \times 8 = 64$

(ii) $\sqrt{81}$, or the square root of 81, means finding the number which, when squared, gives 81. The most obvious answer is 9, because it satisfies this condition – i.e. $9^2 = 81$. However, if you give the question a little further thought you may notice that there is another possible answer, namely, -9. You can check this by squaring -9, thus: $(-9)^2 = -9 \times -9 = 81$

(iii) $\sqrt{(3^2 + 4^2)}$

First, work out what is inside the brackets:

$(3^2 + 4^2) = 9 + 16 = 25$.

Next, find the square root of 25. Using the same reasoning as in part (ii), this gives the two possible solutions, 5 or -5.

QUESTION 2

a $20 + 7 + \frac{3}{10} + \frac{6}{100} + \frac{4}{1000}$

These numbers have been arranged in a familiar pattern – tens, units, tenths, hundredths, and so on. Thus the number can be written down directly as 27.364.

b $60 - 3 + \frac{8}{10} + \frac{2}{100} - \frac{7}{1000}$

This question is similar to part **a**, but slightly complicated by the two values which are subtracted. There is no single correct way of doing this, but my approach was to break it down as follows:

$60 - 3 = 57$

$\frac{2}{100} - \frac{7}{1000} = \frac{20}{1000} - \frac{7}{1000} = \frac{13}{1000} = \frac{1}{100} + \frac{3}{1000}$

Solution: $57 + \frac{8}{10} + \frac{1}{100} + \frac{3}{1000} = 57.813$

c $\frac{62\,341}{1000}$

Division by 1000 has the effect of moving the decimal place three places to the left. The number 62 341 has an invisible decimal point after the 1 (i.e. '62 341.')

Thus, $\frac{62\,341}{1000} = 62.341$

QUESTION 3

The only comment here is that you need to keep in mind the sequence of decimal places, which are as follows:

… thousands hundreds tens units • tenths hundredths thousandths …

QUESTION 4

It is difficult to compare numbers written as fractions and the best strategy is to convert the fractions to either decimals or percentages.

a Converting three-quarters to a percentage:
Three-quarters as a percentage is $\frac{3}{4} \times 100 = 75\%$, which is larger than 70%. Incidentally, if you were unable to find three-quarters of 100 in your head, use your calculator, as follows: Press 3 $\boxed{÷}$ 4 $\boxed{×}$ 100 $\boxed{=}$

b Converting one-twentieth to a decimal:
One-twentieth as a decimal is $\frac{1}{20} = 0.05$, which is smaller than 0.06.
Alternatively, press 1 $\boxed{÷}$ 20 $\boxed{=}$

c Converting two-fifths to a decimal:
Two-fifths as a decimal is $\frac{2}{5} = 0.4$, which is smaller than 0.5.
Alternatively, press 2 $\boxed{÷}$ 5 $\boxed{=}$

d Converting an eighth to a percentage:
An eighth as a percentage is $\frac{1}{8} \times 100 = 12\frac{1}{2}\%$, which is larger than 10%.
Alternatively, press 1 $\boxed{÷}$ 8 $\boxed{×}$ 100 $\boxed{=}$

e Converting a tenth to a percentage:
A tenth as a percentage is $\frac{1}{10} \times 100 = 10\%$, which is larger than 8%.
Alternatively, press 1 $\boxed{÷}$ 10 $\boxed{\times}$ 100 $\boxed{=}$

QUESTION 5

There are no comments on this question except to suggest that you could develop your estimation skills by guessing some measures around the house, then getting out a tape measure, weighing scales, thermometer and so on, and checking how good your guesses were. It is surprising how quickly these skills do improve with practice.

QUESTION 6

As is the case for all questions about comparison of measures, the key thing is to convert all the measurements to the same units. Once this has been done, placing them in order of size becomes a trivial task. Suitable conversions are set out below for the measures, written here in the order in which they were originally given in the question.

a 450 ml; 500 ml; 568 ml; 750 ml
b 175 cm; 18 cm; 120 000 cm; 30 cm
c 84 hours; 96 hours; 95 hours; 87.6 hours.

QUESTION 7

As with most estimation questions, there is no single correct method, as each person draws on their own past knowledge and experience.

a My experience of motorway driving is that traffic on the outside lane seems to travel at around 80 mph (most drivers in the outside lane tend to break the speed limit of 70 mph unless there happens to be a police vehicle in the vicinity). So my first estimate here will be in these imperial units of miles per hour and then I will use my calculator (pressing 70 $\boxed{\times}$ 8 $\boxed{÷}$ 5 $\boxed{=}$) to convert to the metric equivalent, thus:

70 mph = $70 \times \frac{8}{5}$ km/h = 112 km/h. I rounded this to 110 km/h.
90 mph = $90 \times \frac{8}{5}$ km/h = 144 km/h. I rounded this to 145 km/h.

b As with the previous part, I know from experience that 4 mph represents a fairly brisk walk, so I allowed a range of between

3 and 5 mph. The conversions to km/h were done as above in part **a**.

c This time I had no idea how fast a top sprinter could run, so I decided to do a calculation instead. Again, drawing on my past experience, I know that a good time for the 100 m is around 10 seconds. These seemed to be convenient numbers, so I chose to work in metric units this time and converted to imperial afterwards.

In 10 seconds, the sprinter travels 100 m
In 1 minute, the sprinter would travel 100×6 m
In 1 hour, the sprinter would travel
$100 \times 6 \times 60$ m $= \frac{100 \times 6 \times 60}{1000}$ km.
Pressing the calculator sequence 100 $\boxed{\times}$ 6 $\boxed{\times}$ 60 $\boxed{\div}$ 1000 $\boxed{=}$ gives the answer 36. In other words, the sprinter's speed is 36 km/h.

Spotlight: An aside on cancelling out fractions

There is an alternative method of working out this last answer which involves 'cancelling' out the fraction. In general, cancelling out a fraction means dividing numbers in the top and bottom parts of the fraction by factors that they have in common. This has the effect of simplifying the fraction before it is evaluated. In the case of the fraction $\frac{100 \times 6 \times 60}{1000}$, it means dividing out the tens and hundreds. This has been shown in separate stages below.

First, divide the top and bottom of the fraction by 100, giving the following:	$\dfrac{\overset{1}{\cancel{100}} \times 6 \times 60}{\underset{10}{\cancel{1000}}}$
There is still further scope for cancelling, so divide the 60 on the top and the remaining 10 on the bottom by 10, thus:	$\dfrac{\overset{1}{\cancel{100}} \times 6 \times \overset{6}{\cancel{60}}}{\underset{1}{\cancel{10}}}$
This can be tidied up and simplified as follows:	$\dfrac{1 \times 6 \times 6}{1} = 36$

So, the result, as before, is 36 km/h. Again, if you wanted to use your calculator for this calculation, press 100 $\boxed{\times}$ 6 $\boxed{\times}$ 60 $\boxed{\div}$ 1000 $\boxed{=}$.

On the basis of this figure of 36 km/h, I'll allow you to mark yourself correct if your answer falls within the range 30–40 km/h.

(Incidentally, in practice no sprinter could possibly sprint at this speed for an hour, but calculating someone's speed in mph or kmph does not necessarily imply that they continue to travel for an hour, even though my wording above suggests that they do.)

Finally, I can convert to mph as follows.
30 km/h = $30 \times \frac{5}{8}$ mph = 18.75 mph. I rounded this to 20 mph.
Alternatively, press 30 $\boxed{\times}$ 5 $\boxed{\div}$ 8 $\boxed{=}$

40 km/h = $40 \times \frac{5}{8}$ mph = 25 mph, which required no further rounding.
Alternatively, press 40 $\boxed{\times}$ 5 $\boxed{\div}$ 8 $\boxed{=}$

d This again was a fact that I happened to have stored away in my brain. And, having been brought up from childhood with imperial units, I remembered that the speed of sound is around 760 mph. If you had absolutely no idea, try looking up 'speed of sound' in a web search engine. There is no need for any greater accuracy than this because the speed of sound varies, depending on such things as the nature of the gas that it is passing through, the air temperature at the time, and so on. As before, the conversion to km/h is easy with a calculator.
Press 760 $\boxed{\times}$ 8 $\boxed{\div}$ 5 =
I rounded the calculator result of 1216 km/h to 1200 km/h. Supersonic aircraft travel at speeds greater than 760 mph or 1200 km/h. But clearly there needs to be a sensible upper limit to your answer – say, 2000 mph or 3000 km/h.

QUESTION 8
There are no additional comments on this question.

QUESTION 9
There are no additional comments on this question.

QUESTION 10
a The task here is to try to estimate what fraction each slice is of the total. You can see that the slice corresponding to Europe takes up almost half of the pie, so the annual

spending represented by this slice would be almost half of £9.2b, or roughly £4.5b.

b A spending of £2b out of a total spending of £9.2b represents the following fraction of the pie: $\frac{2}{9.2}$.

On my calculator, this gives a decimal value of just over 0.2, or roughly one fifth. So I am looking for a slice which is slightly bigger than one fifth of the pie. Only North America fits the bill here. (If you imagine four more slices the same size as North America, it seems reasonable that five of these slices would together make a complete pie.)

c This final part is an exercise in proportion. We know that 19.2 million visitors spent £9.2b.

So, 1 million visitors should spend $\frac{£9.2b}{19.2}$.

Then 10 million visitors should spend $\frac{£9.2b}{19.2} \times 10$.

On the calculator, press 9.2 ⌶÷⌶ 19.2 × 10 ⌶=⌶ giving the result 4.7916666, which I rounded to £5b.

QUESTION 11

a Since the prices in Hamleys were the highest for each of the three toys listed, this first question was easy to answer.

b There isn't an obvious method for answering this question, but I decided to calculate the total price of all three items and select, as the cheapest, the store with the smallest total, which was Toys Я Us. However, the price differences between John Lewis and Toys Я Us are so small that John Lewis could possibly come out cheapest if three different toys were chosen. Based only on data from three toys, it is impossible to come up with a clear answer to this question.

c and d There are no additional comments on these parts.

QUESTION 12

a There are no additional comments on this part.

b (i) First you must subtract Petra's tax allowances from her annual income. On the calculator, this is done as 6200 ⌶−⌶ 3750 ⌶=⌶ giving the result 2450.

Next multiply the result by 0.2. There is no need to re-enter the 2450 as it is already on the calculator display, so simply press $\boxed{\times}$ 0.2 $\boxed{=}$

(ii) The previous result of £490 should still be on your calculator display. This is the annual tax bill. To calculate this in weekly terms you must divide by 52, so simply press $\boxed{\div}$ 52 $\boxed{=}$, giving the result 9.4230769. I rounded this to the nearest penny, giving £9.42.

c Petra's weekly earnings net of tax can be calculated as follows. Her annual earnings net of tax: Press 6200 $\boxed{-}$ 490 $\boxed{=}$
Her weekly earnings net of tax: Press $\boxed{\div}$ 52 $\boxed{=}$
giving the result 109.80769, which I rounded to the nearest penny, giving £109.81.

Appendix **A**

Calculating a best buy

Whether you are buying potato crisps, rice or shampoo, most supermarket purchases are made available in packs of different sizes and prices. Sometimes you choose the size for practical reasons – here are some examples.

▶ 'The larger toothpaste tubes always fall out of our bathroom mug so I tend to buy the smallest one.'

▶ 'In our household, a small pack of cornflakes lasts about two days so I always buy the largest one.'

But in many situations you simply want to buy the size which gives the best value for money.

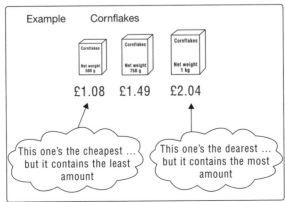

Example   Cornflakes

Cornflakes — Net weight 500 g — £1.08

Cornflakes — Net weight 750 g — £1.49

Cornflakes — Net weight 1 kg — £2.04

This one's the cheapest ... but it contains the least amount

This one's the dearest ... but it contains the most amount

Note: You can't compare the prices directly because each packet contains different amounts of cornflakes. You need to find a way of comparing like with like. There are two possible methods.

For each packet:

A calculate the weight of cornflakes per penny, and then choose the packet which works out at the largest weight per penny

B calculate the cost in pence of cornflakes per gram, and then choose the packet which works out cheapest per gram.

Note: In Method A you end up choosing the packet which produces the largest answer to your calculation, while Method B involves choosing the packet which produces the smallest answer. Remember that your calculator will do the arithmetic. All you have to do is three things, summarized by the letters DPI:

▶ D *Decide* what calculations to do and understand why you are doing them

▶ P *Press* the right buttons in the correct order

▶ I *Interpret* the answers sensibly.

Method A Calculating the weight per penny

D The calculation needed here is division: the weights divided by the price of each packet. To avoid confusion, it makes sense to use the same units for weight and price for each packet. So the weights will be measured in grams and the price in pence.

P The calculation is set out in the table below.

Size	Price (p)	Weight (g)	Grams per p (to 3 figures)
Small	108	500	500 ÷ 108 = 4.63
Medium	149	750	750 ÷ 149 = 5.03
Large	204	1000	1000 ÷ 204 = 4.90

I Choose the size with the largest number of grams per penny, which in this case is the medium packet with 5.03 grams per penny.

Method B Calculating the price per gram

D   As with Method **A**, the calculation required is division, but this time the division is the other way round – price divided by weight.

P   The calculation is set out in the table below.

Size	Price (p)	Weight (g)	Pence per g (to 3 figures)
Small	108	500	108 ÷ 500 = 0.216
Medium	149	750	149 ÷ 750 = 0.199
Large	204	1000	204 ÷ 1000 = 0.204

I   This time we are looking for the size with the cheapest price per gram. As before, we select the medium packet with 0.199 pence per gram.

Points to note in value-for-money calculations

▶   It doesn't matter which method you use (calculating weight per penny or price per gram). You just need to be clear which one you have chosen and make sure to use the same method throughout.

▶   Ensure that the units of measure match up – don't calculate one pack size priced in pence and another in pounds.

▶   Whichever method you adopt determines whether you will choose the packet whose calculation yields the *largest* value or the *smallest* value. Remember that you want to pay small pence and you want to receive large quantities. Thus:

For the calculation ...	you want ...	so you choose...
grams per p	large grams	the largest one
pence per g	small pence	the smallest one

▶   We have only looked at a simple example where the goods being compared were identical in every respect except size and price. For most purchases, there are many other factors to take into account when deciding on value for money and it is altogether more complicated than these calculations suggest. For example, you may also wish to take account of quality, durability, prestige, recyclability, and so on; all factors which are much more difficult to measure and calculate with.

▶ Unit pricing is a practice followed by most supermarkets. As well as including on the label of each item the price and the size (weight or capacity, as appropriate), the 'unit price' is also included. This allows you to compare the relative value of products across different pack sizes. Here are some examples.

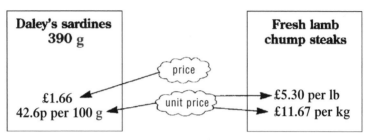

Notice that when supermarkets give a unit price, they usually reduce the price to a *suitable* unit, not necessarily to a single gram or pound. In the examples above, the basic unit for sardines was taken to be 100 g, while that for lamb steaks is both lb and kg. The reason for this is to avoid having to use prices written as awkward decimal numbers that people find hard to make sense of. Note also that the meat has been unit priced in both imperial and metric units for the customer's convenience.

Appendix B

Reading the 24-hour clock

Analogue and digital

The world is divided into two types of device – analogue and digital. Analogue devices are so called because of their mechanical way of working: the mechanism is the device. For example, a vinyl record player produces musical sound in a mechanical way in that, as the record spins, the needle moves about inside the grooves. That movement is then translated into sound. Contrast that with digital sound, where a laser reading device merely scans a long list of numbers (i.e. digits) encoded in the compact disc or digital audio tape. It is these numbers that are then transformed into sound.

a record is analogue

a CD is digital

Another example of this analogue and digital distinction is with clocks and watches.

an analogue clock

$14:35$

a digital clock

The old-fashioned analogue clocks tend to have a circular face numbered 1 to 12 and hands that sweep round, marking out the time. As with the record player, there is a moving mechanism that physically marks out a circular path, which we interpret in terms of time passing. One complete circuit of the clock face by the hour hand represents the passing of 12 hours. Two circuits of the clock face by the hour hand gives a full day of 24 hours. We use common sense to distinguish between morning time (a.m.) and afternoon time (p.m.).

Digital clocks and watches, on the other hand, simply produce numbers. These numbers can be organized in 12-hour cycles, in which case the letters a.m. or p.m. are shown on the display. Alternatively, most digital clocks can be set to display time in cycles of 24 hours. The chart below shows how 12-hour and 24-hour times are related.

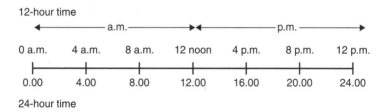

As you can see, for the first 12 hours in a day (i.e. during the a.m. period) the 12-hour and 24-hour times are exactly the same. However, after 12 noon, the times roll back to zero on the 12-hour system, whereas they simply continue (13, 14, 15, ...) on the 24-hour system.

Converting from 12-hour time to 24-hour time

Remember that 24-hour time is a measure of how long it is since the previous midnight. So:

▶ ...if it is a.m., the 24-hour and the 12-hour times are the same, and

▶ ...if it is p.m., you have to add 12 hours.

Here are some examples.

12-hour time	a.m. or p.m.?	+ 12 hours?	24-hour time
3.15 a.m.	a.m.		3.15
6.44 p.m.	p.m.	+ 12.00	18.44
4.52 p.m.	p.m.	+ 12.00	16.52
5.00 a.m.	a.m.		5.00
11.07 p.m.	p.m.	+ 12.00	23.07

Converting from 24-hour time to 12-hour time

Remember that any time after 12 noon is p.m., and for afternoon times the 12-hour clock rolls back to zero. This means that, if the 24-hour time is greater than 12 (i.e. if it is a p.m. time), you must subtract 12 to find the 12-hour time.

Here are some examples.

24-hour time	more than 12?	– 12 hours?	12-hour time
13.55	yes	– 12.00	1.55 p.m.
16.40	yes	– 12.00	4.40 p.m.
4.08	no		4.08 a.m.
11.50	no		11.50 a.m.
21.33	yes	– 12.00	9.33 p.m.

Finally, here are some diagrams to help you sort out what to do when converting between 12-hour and 24-hour time.

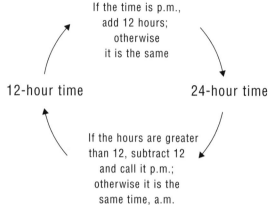

If the time is p.m., add 12 hours; otherwise it is the same

12-hour time

24-hour time

If the hours are greater than 12, subtract 12 and call it p.m.; otherwise it is the same time, a.m.

From 12-hour to 24-hour time

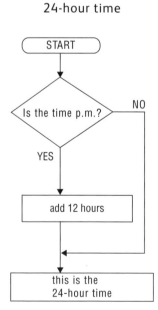

START

Is the time p.m.? NO

YES

add 12 hours

this is the 24-hour time

From 24-hour to 12-hour time

START

Is the time more than 12.00? NO

YES

subtract 12 hours

this is the 12-hour time

Why?

The chief virtue of the 24-hour system is that it automatically does away with the need to specify whether the time is a.m. or p.m. For most everyday purposes, this may not seem very important, but when consulting train, bus and airline timetables, it makes sense to use a system which is not prone to confusion. It has been estimated that, over the first ten years of the introduction of the 24-hour clock in their timetables, rail staff costs fell by nearly £8 million in today's terms, due to dispensing with the need to pursue interminable conversations with customers along the lines of, 'Excuse me. Is that 4 o'clock a.m. or 4 o'clock p.m.?' etc.

(Actually, I just made that last 'fact' up, but you get the general point!)

Appendix C

Bus and railway timetables

Bus, railway and aeroplane timetables are invariably written in terms of the 24-hour clock. Before proceeding with this appendix, make sure you understand and can use the 24-hour clock (see Appendix B).

Two extracts from a railway timetable are shown here. Notice that both timetables are labelled Table 5, but each one has a network map above it showing the direction of travel. The second timetable shows the journey *to* London Euston (which will be the outward journey for our purposes), while the first timetable covers the return journey *from* Euston station.

It is a Tuesday morning and you are in Wilmslow. You have arranged to meet a friend in a cafe in London at 12 noon. You should allow about half an hour to travel by tube from Euston station to get to the cafe. You want to be home by 7 p.m. that evening (you live about 45 minutes from Wilmslow station).

Now try to answer the following questions.

a What train should you catch if you want to be certain of arriving at the cafe before your friend? How long are you likely to have to wait at the cafe if you catch this train?

b What is the most sensible train to catch?

Table 5 INTERCITY West Coast

Full service London — Rugby Table 4.   Full service London — Crewe, Table 6.

Mondays to Fridays	🚉☒	🚉🅴Ⓢ A	🚉✗	🚉✗	🚉✗	🚉✗	🚉✗	🚉✗	🚉✗	🚉✗
London Euston	0655	0800	0900	0910	1000	1020	1100	1200	1300	1310
Watford Junction	0711u	0816u	——	0926u	1016u	1037u	——	1216u	1241	1326u
Milton Keynes Central	0734	——	0936	——	——	——	1135	——	1334	——
Rugby	——	——	——	1011	——	——	——	1301	——	1411
Stoke-on-Trent	0844	——	1046	1130g	1144	——	1244	——	1444	1531g
Macclesfield	0903	——	1105	1152g	1203	——	1303	——	1503	1549g
Crewe	——	——	——	1112	——	1222	——	——	——	1512
Wilmslow	——	1010	——	1204	——	1304	——	1414	——	1604
Stockport	0918	1020	1120	1207g	1218	——	1318	1424	1518	1605g
Manchester Piccadilly ↓	0932	1033	1132	1219g	1234	——	1332	1437	1532	1619g

Mondays to Fridays	🚉✗	🚉✗	🚉✗	🚉🅴Ⓢ A	🚉✗	🚉🅴Ⓢ A	🚉	🚉🅴Ⓢ A	🚉✗	🚉✗
London Euston	1400	1410	1500	1600	1630	1700	1735	1800	1805	1900
Watford Junction	1416u	——	——	1616u	1646u	——	——	1816u	1822u	——
Milton Keynes Central	——	1444	1534	——	——	——	——	1845u	——	——
Rugby	——	1511	——	——	——	1834	——	1912	——	——
Stoke-on-Trent	1544	1619g	1644	——	1814	——	1938	1950	——	2045
Macclesfield	1603	——	1703	——	1833	——	1957	2009	——	2108
Crewe	——	1612	——	——	——	——	——	——	2008	——
Wilmslow	——	1704	——	1808	——	1913	——	——	2104	——
Stockport	1618	1716	1718	1818	1848	1923	2012	2024	2116	2123
Manchester Piccadilly ↓	1636	1728	1732	1834	1902	1936	2025	2037	2128	2137

Mondays to Fridays	🚉✗	🚉	Saturdays							
London Euston	2000	2200	——	——	0645	0730	0750	0800	0850	0900
Watford Junction	1416u	2216u	——	——	0701u	——	0808u	——	0824	0916u
Milton Keynes Central	2034	2239	——	——	0724	0805	——	0834	0924	——
Rugby	——	——	——	——	——	0844	——	——	——	1011
Stoke-on-Trent	2144	——	——	——	0844	——	0944	——	1046	1130g
Macclesfield	2203	——	——	——	0903	——	1003	——	1105	1152g
Crewe	——	0020	——	——	——	0942	——	1012	——	1112
Wilmslow	——	0042s	——	——	——	1016	——	1104	——	1204
Stockport	2218	0102s	——	——	0918	——	1018	——	1120	1207g
Manchester Piccadilly ↓	2232	0114	——	——	0932	——	1034	——	1133	1219g

Saturdays	🚉	🚉	🚉
London Euston	0950	1000	1050
Watford Junction	1008u	——	——
Milton Keynes Central	——	1034	1125
Rugby	——	1111	——
Stoke-on-Trent	1144	——	1244
Macclesfield	1203	——	1303
Crewe	——	1212	——
Wilmslow	——	1304	——
Stockport	1218	1316	1318
Manchester Piccadilly ↓	1234	1329	1332

On-Board Services (see page 1):
🚉 InterCity train with catering.
🅱 First Class Pullman.
✗ Restaurant.
🔲 Restaurant (First Class only).
Ⓢ Silver Standard.
🔲 Seat reservations essential free of charge.

Light printed timings indicate connecting service.

Notes for this and opposite page:
A   The Manchester Pullman.
C   Until January 8.
D   Until Jan 8; then from Apr 16.
E   Until Jan 7/8; also May 27/28.
G   Jan 14/15 to May 20/21.
H   From January 15.
J   January 15 to April 9.
g   Change Stafford.
k   0044 May 28 only (Sunday morning).
n   0120 May 28 only (Sunday morning).
p   Until Jan 8; then from Apr 16 at this station.
q   Jan 15 to Apr 9 at this station by special bus from Crewe.
s   Calls to set down only.
u   Calls to pick up only.

Table 5 INTERCITY West Coast

Manchester Piccadilly — Stockport — Macclesfield — Stoke-on-Trent — London Euston (via Crewe — Watford)

Full service Crewe — London, Table 6. Full service Rugby — London, Table 4.

Mondays to Fridays

Station			(A)	(A)						
Manchester Piccadilly	0520	0600	0645	0710	0730	0830	0833	0930	1030	1033
Stockport	0528	0608	0654	0718	0738	0838	0842	0938	1038	1042
Wilmslow	0536	——	——	0727	——	——	0850	——	1046	1050
Crewe	0557	——	——	——	——	——	0925	——	——	1125
Macclesfield	——	0622	0706	——	0752	0852	——	0952	——	1039g
Stoke-on-Trent	——	0642	0727	——	0812	0912	——	1012	——	1102p
Rugby	——	0730	——	——	——	——	1023	——	——	1223
Milton Keynes Central	——	0824	——	0924	——	——	1049	——	1221	1247
Watford Junction	——	0856	——	0921s	1013	1045s	1129	1142s	1313	1329
London Euston	0813	0840	0916	0944	1011	1108	1136	1205	1307	1334

Mondays to Fridays

Station									(A)	
Manchester Piccadilly	1130	1133	1230	——	1330	1333	1430	1433	1530	——
Stockport	1138	1142	1238	——	1338	1342	1438	1442	1538	——
Wilmslow	——	1150	——	1250	——	1350	——	1450	——	1527
Crewe	——	1225	——	1325	——	1425	——	1525	——	1608
Macclesfield	1152	——	1252	——	1352	——	1452	——	1552	——
Stoke-on-Trent	1212	——	1312	——	1412	——	1512	——	1612	——
Rugby	——	1325	1359	——	——	1527	——	1628	——	——
Milton Keynes Central	——	——	1424	1447	——	——	1618	1653	——	——
Watford Junction	1342s	1412s	1513	1529	1542s	1614s	1659	——	1742s	1753s
London Euston	1405	1435	1510	1534	1605	1637	1705	1739	1805	1816

Mondays to Fridays

Station	(A)							
Manchester Piccadilly	1630	——	1730	——	1830	——	1933	2000
Stockport	1638	——	1738	——	1838	——	1942	2008
Wilmslow	1648	——	——	1734	——	1825	1950	——
Crewe	——	1723	——	1823	——	1915	2026	——
Macclesfield	——	1640g	1752	——	1852	——	2022	——
Stoke-on-Trent	——	1700g	1812	——	1912	——	2042	——
Rugby	——	1823	——	1921	——	——	2134	——
Milton Keynes Central	1820	1847	1918	——	2018	——	2159	2204
Watford Junction	1856	——	1947s	2010s	2113	——	2227s	2232s
London Euston	1905	1934	2010	2033	2105	2114	2250	2305

Saturdays

Manchester Piccadilly	0530	0630	0700	0730	0830	0833	0930	1030	1033	1130
Stockport	0538	0638	0708	0738	0838	0842	0938	1038	1042	1138
Wilmslow	——	——	0716	——	——	0850	——	1046	1050	——
Crewe	——	——	0736	——	——	0925	——	——	1125	——
Macclesfield	0552	0652	——	0752	0852	——	0952	——	1039h	1152
Stoke-on-Trent	0612	0712	——	0812	0912	——	1012	——	1102h	1212
Rugby	——	——	——	——	——	1023	——	——	1223	——
Milton Keynes Central	——	0820	——	0924	——	1049	——	1221	1247	——
Watford Junction	0755s	0856	0923s	——	1045s	1129	1142s	1313	1329	1342s
London Euston	0828	0917	0956	1021	1118	1146	1215	1317	1344	1415

Notes for this and opposite page:
- **A** *The Manchester Pullman.*
- **g** Change Stafford.
- **h** Change Stafford (May 27 only Macclesfield 1037 Stoke-on-Trent 1056).
- **p** Until January 8 at this station by special bus to Crewe.
- **q** 1752 Jan 15 to Apr 9 by special bus to Crewe.
- **s** Calls to set down only.
- **u** Calls to pick up only.

On-Board Services (see page 1).
- InterCity train with catering.
- First Class Pullman.
- Restaurant.
- Restaurant (First Class only).
- Silver Standard.
- Seat reservations essential free of charge.

Light printed timings indicate connecting service.

> Is this a direct service or will you have to change trains?

> If you have to change, where are you likely to have to change and how long may you have to wait for that connection?

> At what time would you estimate arriving at the cafe?

c What train will you catch to return home? Will you be able to eat on this train?

d What are the train journey times each way?

SOLUTION

a For the outward journey, you need the second timetable. The 0727, which gets into London Euston at 0944, should get you to the cafe by 1015 – an hour and three-quarters before the agreed time, so not very satisfactory!

b It's a bit tight, but you should just about make your assignation if you catch the next train from Wilmslow, the 0850, getting into London Euston at 1136.

> The 0850 departure is not a direct service. You can tell this because of the light printing of the departure time of 0850. As you can see from the explanation at the bottom of the page, 'Light printed timings indicate connecting service', so this will require a change of trains.

> In this case you will have to change at Crewe, which is the next main station. Note that the times 0850 (in light printing) and 0925 (in bold) are the *departure* times from Wilmslow and Crewe, respectively. Since you will be changing at Crewe, you will expect to arrive there some time before the 0925 departs. You can make an intelligent guess at your arrival time in Crewe by looking at a previous column of figures. Notice that the 0536 from Wilmslow gets into Crewe at 0557, suggesting a journey time of 21 minutes. Assuming the 0850 travels at the same speed, it should get into Crewe by 0911, allowing you ample time (14 minutes, in fact) to make your connection on the 0925.

> This should get you to the cafe just a few minutes after 12 noon.

c To choose the homeward train, you need to work backwards from when you want to get home, as follows:

Getting home by 7 p.m. means arriving at Wilmslow station by 6.15 p.m., i.e. by 1815. According to the first timetable, there is an ideal train which departs from London Euston at 1600 and gets into Wilmslow at 1808. *Note*: According to the codes at the top of the column, this train is a First Class Pullman (called *The Manchester Pullman* – see note A on the timetable) with a Silver Standard, but with no Restaurant facilities. So, you will be able to eat on this train, but not in high style!

d The journey times are shown in the table below

Depart		Arrive		Journey time
Wilmslow	0850	London Euston	1136	2 hours 46 mins
London Euston	1600	Wilmslow	1808	2 hours 08 mins

So, the return journey is quicker by some 38 minutes.

You may have gone wrong calculating journey times on your calculator. For example, the outward journey ran from 0850 to 1136, so pressing 1136 ⊟ 0850 ⊒ gives 286 (i.e. 2 hours and 86 mins) and not 246 (or 2 hours 46 mins) as shown above.

Calculating journey times cannot easily be done on a calculator. The complication is that there are 60, not 100, minutes in one hour. There are many commonsense ways of tackling this problem. Here is how I worked out the journey time between 0850 and 1136.

▷ First, I added ten minutes to 0850 to bring it up to the next exact hour (0900). This shortens the journey time by ten minutes, but I'll add these on later.

▷ Next, I calculated the journey time from 0900 to 1136 – this is easy to do in your head, the answer being 2 hours 36 minutes.

▷ Finally, I need to remember that I shortened the journey time by ten minutes, so I must now add these on.

So, 2 hours 36 mins + 10 mins gives the answer, 2 hours 46 minutes.

Appendix **D**

Checking the supermarket bill

Most people simply haven't got the time or the energy to check their weekly supermarket bill item by item. In general, we tend to assume that the machine has got it right. Where errors occur, sometimes they are to the customer's advantage and sometimes to the store's advantage, but it is likely that, taken over the long term, errors tend to average themselves out.

Typically, the item will be scanned at the checkout with a barcode reader. Occasionally the barcode is read incorrectly, but barcodes have a built-in 'checksum' that bleeps when there is an error – this is explained in Appendix K. However, errors do occur, and it is worth being awake to that possibility at the checkout, to avoid it costing you money. The key thing is to know when to check the bill in detail and how to do so if you have to. Here are a few guidelines.

What does my bill usually come to?

First of all, it is helpful to know roughly what your bill is likely to come to. If you do a regular weekly shopping at the same store, you may be able to do this with some accuracy. For example, your typical bill may come to, say, around £55, so anything less than £40, or more than £70, in a particular week ought to make you suspicious.

Are there any unusual and expensive items this week?

If you have bought some untypical and expensive items (for example, alcohol, or kitchen or electrical goods), make an estimate of these and add them to your typical bill. That way you will know what you can expect your bill to come to, roughly.

How many items did I buy?

Most supermarket receipts state the number of goods bought. In a recent large shop for groceries, I spent about £90, having bought 51 items. This works out at just under £2 per item.

P PERKINS PLC	
THE BUTTS	
WARWICK	
CV21 3FL	
Telephone no. 01926 334215	
	£
Custard powder	1.27
Apple jce 2L	1.94
PP County spread	2.85
PP H-gran stick	0.74
Baking potatoes	2.28
PP Earl Grey tea	1.89
PP Earl Grey tea	1.89
PP Lentils 500g	1.38
Orange jce 1L	2.48
PP Eggs large	2.35
Baked beans	0.82
11 Bal Due	**19.89**
4356 601364 645355431	

For most supermarket bills, an average of just under £2 per item is fairly typical, and using this fact might provide you with a quick check that the overall bill is in line with the contents of your trolley. Thus, buying, say, 30 items, I might expect the bill to be around £55 – £60. This strategy might allow you to pick up situations where, for example, £69 was recorded instead of 69p for a tin of beans.

Can I do a quick mental check of the bill?

If there is a huge number of items on your bill – as many as 73, for example – it isn't realistic to do a mental check. However, if there are, say, only 11 or 12 items, then this is perfectly possible. There is no single correct method of rounding – you could round to the nearest 10 pence or 50 pence. The method below is more approximate and is based on rounding the prices to the nearest £.

The method is based on looking at the pence part of the price. Where the amount of the pence is 50p or more, round up to the next whole number of pounds; otherwise ignore the pence, which has the effect of rounding down the pounds. For example, the first item, which is 85p, will be rounded up to £1 because the pence (85p) is greater than 50p. On the other hand, a sum of £1.19 is rounded down to £1 because the pence (19p) is less than 50p.

Adopting this procedure, the costs of the 11 items are rounded as follows:

Item	Actual price (£)		Rounded price (£)
Custard powder	1.27	→	1
Apple jce 2L	1.94	→	2
PP County spread	2.85	→	3
PP H-gran stick	0.74	→	1
Baking potatoes	2.28	→	2
PP Earl Grey tea	1.89	→	2

Continued

Item	Actual price (£)		Rounded price (£)
PP Earl Grey tea	1.89	→	2
PP Lentils 500g	1.38	→	1
Orange jce 1L	2.48	→	2
PP Eggs large	2.35	→	2
Baked beans	0.82	→	1
Rounded total			£19

In this particular case, the rounding method has produced an answer which is slightly too low – by roughly £1. However, the object of the exercise has been to produce a rough 'order of magnitude' answer, to check whether the total is more or less correct, rather than to get a precise answer.

Appendix E

Understanding a shop receipt

If you've ever fancied yourself as a latter-day Poirot or Sherlock Holmes, you could do a lot worse than to practise your skills uncovering the hidden mysteries of a lowly shop receipt. You may be surprised to discover how much you can tell about a person, simply by rummaging around in their discarded plastic bags and fishing out the sordid details of their last shopping transaction – which might look something like the receipt below...

```
                    S DEVLIN PLC
                  THE SHIRES PARK
                   DODDINGTON
                     CV11 4RR
             TELEPHONE NO. 01926 435268
         SALES VOUCHER: CUSTOMER'S COPY
0873456        D/WASH LIQ LEM           3.29
0786543        TULIPS                   2.95
2 BAL DUE                               6.24
CASH                                   10.00
CHANGE                                  3.76
     503 24 1032        16:26       24DEC08
             VAT NO. 534 5644 89
     THANK YOU FOR SHOPPING WITH DEVLIN
           Please retain your receipt
```

As you can see, this piece of evidence tells its own story about the transaction that took place. You might like to reconstruct part of that story, sorting out in your mind the objective facts involved – the 'where', 'when' and 'what' of the transaction. To help you, try completing the blanks in the police file below.

Police file

We have reason to believe that the suspect entered the premises of _____ (shop) at _____ (address) in the town of _____ on the afternoon of _____ in the year _____. At precisely _____, _____items were purchased, namely a _____ and a _____, costing £_____and £_____respectively. A £_____ note was submitted to the cashier and £_____ in change was received.

For the completed file see the end of Appendix E.

We might like to go a little further here and speculate what sort of person this was. Note the date, which was Christmas Eve. Now most people who celebrate this festival are still frantically buying the basics on Christmas Eve (in a few homes the turkey, crackers, balloons, presents, etc. have still to be bought). This individual has clearly got the whole Christmas thing totally under control, if he or she is making a special trip for tulips and dishwasher liquid on Christmas Eve! Also, assuming that they own the dishwasher in question, you might suppose that they aren't exactly in the bottom income bracket.

In short, one has a picture of someone who is reasonably well off, certainly well organized, who wants to relax this Christmas with no plans to be hand-washing dishes in the sink!

Follow up

You might like to find various tickets and receipts of your own and see if you can crack all the codes they contain.
I have focused on what information they provide for the customer. What additional information do they provide for the management? How might they be used for stock control?

Appendix F

Checking the VAT

Value added tax (VAT) is charged on many of the goods bought in the UK. In 2011, the rate of the tax was raised from 17.5% to 20%. What this means is that, for every £100 net value, the VAT charge is £20, bringing the total to £120. In other words:

Net value + VAT   = Gross value
£100.00   + £20.00 = £120.00

This is what the customer pays

Below is a simplified receipt from a plumbing centre where I recently bought the wherewithal to install a ventilator into an external wall of my kitchen. (In the event I failed disastrously to complete the job without professional help, but that's another story!)

Catalogue No.	Qty. Ordered	Description	VAT	Price	Total net
800160	1	Stadium BM720 ventilator	20	£22.40	£22.40
602047	1	Geocel Fast Set cement – 3kg	20	£7.16	£7.16
				Subtotal	£29.56
				VAT (20%)	£5.91
				Total	£35.47

Let's check that the basic arithmetic is correct.

a Is the subtotal of £29.56 correct?

From the right-hand column, we can see that the two items cost £22.40 and £7.16, respectively. Pressing 22.40 ⊞ 7.16 ⊟ on the calculator confirms the answer given in the subtotal box, £29.56.

b Is the VAT of £5.91 correct?

Notice that in this receipt the net totals are added first, and then the overall VAT is calculated at the end. In order to check the VAT, you must find 20% of 29.56. As was explained in Chapter 6, this is found by converting 20% to its decimal form (giving 0.20, or simply 0.2) and multiplying this by 29.56, thus:

0.2 ☒ 29.56 ⊟, giving an answer 5.91.

c Is the total of £35.47 correct?

Adding the net subtotal and the VAT should give the overall gross total, thus: 29.56 ⊞ 5.91 ⊟, confirming the final bill of £35.47.

Some additional questions

> Surely it's not necessary to work out the VAT on its own if I simply want to check that the overall bill, inclusive of VAT, is correct?

You are quite correct – it is not necessary to find the VAT first and then add it on! The VAT-inclusive bill can be found directly by multiplying the net bill by a suitable multiplying factor. To increase by 20%, multiply by 1.20, or simply 1.2, thus:

1.2 ☒ 29.56 ⊟

This gives 35.472, so after rounding, this calculation confirms the final bill of £35.47.

> My calculator has a percentage key marked on one of the buttons. How can I use it to work out VAT?

Unfortunately not all calculator percentage keys are designed to work in the same way, so you may need to consult your calculator manual to check this for yourself. However, here are some suggestions for things to try. Incidentally, if you are trying things out on a calculator to see how it works, choose simple numbers! I suggest that you try to add, say, 8% on to 100, knowing that the answer should be 108.

Try pressing this sequence and see what you get:

100 [+] 8 [%]

(Note that you may need to complete the calculation with [=])

Similarly, to calculate a percentage reduction of, say, 8%, try pressing:

100 [−] 8 [%]

(with or without the final [=])

On some calculators you will simply not get a satisfactory result to this calculation. As an aside, another use of the [%] on some calculators is converting fractions to percentages, thus:

3 [÷] 5 [%] may produce the answer 60, because the fraction $\frac{3}{5} = 60\%$. Other calculators give weird answers to this key sequence, so a sensible strategy is to experiment for yourself and get to know the quirks of your particular calculator.

Overall, then, the percentage key is a bit of a mixed blessing! It may be useful in VAT calculations, but, provided you understand how to convert a percentage to a decimal, you really don't need such a key.

Appendix **G**

Cooking with figures

> *As a child living in Ireland, I remember watching my grandmother baking soda bread on a griddle. I asked her how she did it.*
>
> *'Well,' she said, 'you start by taking two gopins of flour ...'*
> *She then had to explain to me that a 'gopin' was a double handful.*
> *'But how do you know when you've got exactly a gopin?' I asked.*
> *'Oh, you just know by the feel of your hand,' she replied.*

Measurements in cooking these days tend to be much more precise. Recipes are usually stated in formal units like grams, lbs, litres, and so on; these were explained in Chapter 7, Measuring. This section covers one or two specific questions that often present themselves in the kitchen, which require some mathematics.

How do teaspoons, pints and litres match up?

Most recipes published in the UK tend to be stated in both imperial and metric units, as well as in more informal units such as teaspoons, tablespoons, drops, etc. The imperial measures are based on British weights and liquid measures. Note that American measures are different. For example, a standard American cup will hold 4 oz of sifted flour, as compared with a standard British cup of 5 oz. Similarly, there are roughly 3 British tablespoons of sifted flour to the ounce, as compared with 4 American tablespoons to the ounce.

The table below summarizes the approximate capacities of the informal measures around the kitchen.

1 teaspoon	= 5 ml		
3 teaspoons	= 1 tablespoon (tbsp)		
1 tablespoonful	= 15 ml		
1 teacupful	= $\frac{1}{3}$ pint	= 7 fluid ounces	= 190 ml
1 breakfastcupful	= $\frac{1}{2}$ pint	= 10 fluid ounces	= 280 ml

There is no exact whole number conversion between metric and imperial measures, so whatever value you choose will depend on how accurate you need to be. In cooking, the needs for accuracy are usually not great, and indeed you wouldn't be able to weigh out ingredients to great accuracy anyway. The tables below give accurate and approximate conversions between imperial and metric measures.

Weight

	Multiply by	
To convert	Accurate figure	Cooking approximation
Ounces to grams	28.350	25
Pounds to grams	453.592	450
Pounds to kilograms	0.4536	0.45
Grams to ounces	0.0353	0.035
Grams to pounds	0.0022	0.0022
Kilograms to pounds	2.2046	2.2

Liquid measures

	Multiply by	
To convert	Accurate figure	Cooking approximation
Pints to millilitres (ml)	568	550
Pints to litres (l)	0.568	0.55
Fluid ounces to ml	28.4	25
Fluid ounces to litres	0.0284	0.025
Millilitres to pints	0.00176	0.0017
Litres to pints	1.760	1.75
Millilitres to fluid ounces	0.0352	0.035
Litres to fluid ounces	35.21	35

How accurate do I need to be in my cooking?

This is a difficult question to answer precisely. With some recipes, for example vegetable soup or a salad mix, it isn't critical if you don't use the exact proportions stated in the recipe book. But if you are making, say, a subtle sauce (creamy paprika dressing, for example) the flavour could be affected by even a small error in one of the ingredients.

Have a look now at the basic ingredients for bread and butter pudding, as given in my recipe book, and see if you can spot some sources of error in the measurement of these ingredients.

Bread and butter pudding

Thin slices of wholemeal bread 4 (about 4 oz/100 g)
Butter or margarine 1 oz (25 g)
Raw brown sugar 1 tbsp (15 ml)
Mixed sultanas, raisins & currants 2 oz (50 g)
Fresh milk $\frac{3}{4}$ pt (426 ml)
Free-range eggs 2
Ground cinnamon $\frac{1}{4}$ tsp (1.25 ml)
Nutmeg $\frac{1}{4}$ tsp (1.25 ml)
Serves 4

Here are a few points to note.

a Certain tiny amounts, like the 1.25 ml of nutmeg and ground cinnamon, are too small to weigh on kitchen scales. So you really will need to resort to using the informal measure of $\frac{1}{4}$ tsp. It is actually unclear what this looks like. Recipes sometimes talk about a 'flat teaspoonful' and a 'heaped teaspoonful', so an ordinary teaspoonful is somewhere between the two. Measuring out a quarter of one of those is no easy task. The truth is that this sort of measure is very approximate indeed; and cooks will put in a variable amount of cinnamon and nutmeg, depending on whether or not they are keen on these flavours in their bread and butter pudding.

b Butter and margarine are rarely weighed out. Apart from the fact that they are difficult to weigh out, as they tend to smear the weighing pan, it isn't necessary to do so. The standard method is to take a fresh pack of butter or margarine, which weighs, say, 500 g, and mark it out into five equal sections, thus:

50 g				
25 g				100 g
25 g				

Each main section will therefore be 100 g. Then take half and half again of one 100 g strip, and this is 25 g.

c The amount of egg in the pudding will depend on the size of eggs used and egg size is not specified in the recipe. There is a considerable variation in egg weight, from 'very large' (73 g and over) down to 'small' (53 g and under). Egg sizes are classified into four weight bands, as follows.

Size	Weight
Very large	73 g and over
Large	63 – 73 g
Medium	53 – 63 g
Small	53 g and under

d If you assume that a given 'very large' egg weighs 75 g and a given 'small' egg weighs 50 g, there is 50% more in the 'very large' egg than in the 'small' egg. Looking at it another way, three 'small' eggs weigh roughly the same as two 'very large' eggs.

e Finally, have a look at the imperial and metric measures in this recipe (and others in your own recipe book). As was explained earlier, the conversions are only approximate. But just how approximate are they? The answer is that some are more approximate than others. It is possible to calculate the percentage error of the conversions and this is shown in the table below.

Imperial	Stated metric	Actual metric	Error	Error (%)
4 oz	100 g	113.4 g	13.4 g	$\frac{13.4}{113.4} \times 100 = 12\%$
1 oz	25 g	28.35 g	3.35 g	12%
$\frac{3}{4}$ pt	426 ml	426 ml	0 ml	0%
$\frac{1}{2}$ pt	300 ml	284 ml	16 ml	6%
1 lb	450 g	453.592 g	3.592 g	0.8%

As is clear from the table, some conversions contain a substantial error (for example, the standard conversion from ounces to grams is 12% out) while others, like the $\frac{3}{4}$ pt of milk, contain no evidence of error. Of course, whether you are able, accurately, to measure out exactly 426 ml of milk in your measuring jug is another question!

How do I scale up a recipe?

The recipe for bread and butter pudding given earlier serves four people, but I often cook for seven. This requires having to multiply each amount by the fraction $\frac{7}{4}$. The easiest way to do this is to set the calculator constant to $\times 1.75$. (Using the calculator constant was explained in Chapter 3.) The results can then be rounded *sensibly*.

Original recipe	Scaled up	Rounded
Thin slices of wholemeal bread 4 (about 4 oz/100 g)	$100 \times 1.75 = 175$ g or $\frac{7}{4}$ $\times 4$ (original slices)	175 g 7 slices
Butter or margarine 1 oz (25 g)	$25 \times 1.75 = 43.75$ g	50 g
Raw brown sugar 1 tbsp (15 ml)	$15 \times 1.75 = 26.25$ ml	25 ml
Mixed sultanas, raisins and currants 2 oz (50 g)	$50 \times 1.75 = 87.5$ g	100 g
Fresh milk $\frac{3}{4}$ pt (426 ml)	$426 \times 1.75 = 745.5$ ml	750 ml
Free-range eggs 2	$2 \times 1.75 = 3.5$	4 small/ 3 large
Ground cinnamon $\frac{1}{4}$ tsp (1.25 ml)	$\frac{1}{4} \times \frac{7}{4} = \frac{7}{16}$ tsp	$\frac{1}{2}$ tsp
Nutmeg $\frac{1}{4}$ tsp (1.25 ml)	$\frac{1}{4} \times \frac{7}{4} = \frac{7}{16}$ tsp	$\frac{1}{2}$ tsp

Note: Sensible rounding of the larger metric numbers means rounding to the nearest 25 g or 25 ml. With ingredients like eggs, you can't easily add fractions of an egg, but you may have some flexibility over the size of eggs you use – for example, in this case, 3.5 eggs may approximate to either 4 small eggs or 3 large ones. But if you are like me, you just have one size of egg in your fridge, and so you are stuck with what you've got!

Appendix **H**

Buying a TV set

Something like 96 per cent of households in the UK have at least one television set. Each of these households has therefore taken a decision about whether to buy or rent. If they chose to buy, they had a further choice as to whether to pay it all off straight away, or to put down a deposit followed by regular instalments. The instalment method is also known as 'buying on credit' or HP (hire purchase). This method of payment is a bit like taking out a loan, and you should expect to be charged more for paying in this way than for buying your TV outright.

This example focuses on how much you are likely to pay for your TV set if you decide to 'buy on credit'.

0% finance

Some shops offer a deal whereby you can buy on credit, but the amount you pay overall is the same as if you bought the item outright. This will be advertised as 0% finance or 0% interest. For example:

SONY BRAVIA KDL32RD433BU 32"
LED Television

▶ 81 cm visible screen size   ▶ Tuner: Freeview HD

▶ HD Ready 720p   ▶ Connectivity: HDMI × 2

18 months 0% interest
Price £349.99
20% deposit & 18 direct debit monthly payments of £15.56

Before checking the interest payments, let's just consider how the 'size' of this television set has been described in the advertisement. It is given separately in both imperial units (32 inches) and metric units (81 cm). Measurements stating the size of a television set refer to the length of the diagonal of the screen, measured from corner to corner. Now, let's confirm that 32 inches and 81 cm really are the same length. To convert from inches to cm, multiply by 2.54, thus:

$$32 \times 2.54 = 81.28$$

Rounding down, this gives 81 cm, which confirms the metric length stated in the advertisement.

Something else worth checking here is the claim that this method of payment by monthly instalments really does represent 0% interest.

The figures can be checked as follows:

First, for convenience, let's round up the price of the TV set to £350.

Deposit = 20% of £350	= 0.2 × 350	= £70
18 monthly payments of £15.56	= 18 × 15.56	= £280.08
	TOTAL	= £350.08

OK, so you pay an extra 9 pence (£350.08, as compared with £349.99), but basically the total amount paid out by the instalment method is the same as the cash price. This confirms the claim that this method of payment does represent 0% interest.

By the way, don't assume that the 0% interest deal is always the best. Stores offering such deals may actually have higher prices for similar products than their rivals, who may be offering a higher interest rate. In other words, the cost of the loan may be included in the price.

APR

0% interest is good when you can get it, but usually there is some interest charge when paying on credit. It is useful to know exactly how much you are being charged, and to be able to compare the 'real' interest rate between different shops. Dealers charge a variety of different interest rates, subject to the size of the deposit and the length of the repayment period. As a result, it can be difficult to compare the actual interest being applied from one dealer to another. In recent years, this problem has been solved by the fact that all retailers are legally required to publish the effective interest rate of each deal on offer, using a measure called the 'annual percentage rate' or APR. The APR is the percentage cost of the loan, calculated over a year. It is quite difficult to calculate, as the buyer pays a bit back at a time. The main thing to remember about APR is that a higher rate means that you pay more. For example, an APR of 32% means that you pay out more than with an APR of 27%. In general, all other things being equal (such as price, quality, after-sales service, insurance, and so on), look for the deal offering the lowest APR.

Appendix

Will it fit?

Purchases of large household items, like a sofa, cabinet, or kitchen unit, are often closely linked to the questions 'Where will I put it?' and 'Will it fit?'. Ideally, these questions are sorted out *before* you have parted with your money, and not after!

Moving house is another situation where questions of arranging the furniture have to be made sensibly and, ideally, made in advance of removal day.

A useful strategy for helping you to decide where items of furniture should go is to produce a scale drawing of the various rooms, and to make paper cutout models of the sofa, table, TV, shelving unit, etc. This enables you to try things out without any of the sweat of doing it with the objects themselves *in situ*.

The task of producing scale models and a scale drawing is actually quite straightforward and fun to do.

You will need the following resources:

▶ several sheets of squared paper or graph paper

▶ scissors

▶ tape measure

▶ ruler, pencil and access to the back-of-an-envelope!

Then, follow the steps below.

1 Draw a rough 'back-of-an-envelope' sketch of the room, marking on the significant features – windows, doors, chimney breast, etc., and make a note of 'fixed points' like electric sockets, TV aerial, etc.

2 Using a tape measure, measure all the lengths of the room or rooms that you think you will need for your scale drawing. You are recommended to use metric units (metres and centimetres), as these are easier to deal with on the drawing.

3 Get some squared paper or graph paper and decide on a suitable scale. Ideally, the final drawing should take up most of this sheet of paper (if the drawing is too small, you won't have much confidence in decisions where the fit is rather tight). Make the scale drawing on the squared or graph paper.

4 Measure the length and the width of each major item of furniture that might go in the room. Make a 2D drawing, to scale, of each item on another sheet of squared or graph paper. Write the name of each item on its appropriate scale drawing, and then cut the models out.

You are now ready to 'play'!

Here is how I went about it for one room in my 'des. res.', the bedroom.

1 & 2 I made a sketch of my bedroom, measured the dimensions and marked them on, as shown below.

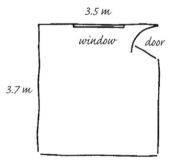

3 My squared paper is marked out in half centimetre squares. It is roughly 60 squares long and 40 squares wide. Since the

bedroom is 3.7 m long and 3.5 m wide, I need to make a sensible judgement about the scale. (This is the only slightly tricky part of the job.) I decided to let 1 m = 10 squares. This meant that the room would be contained in a drawing of 37 squares by 35 squares.

4 The bedroom furniture was duly measured and again the same scale was applied. For example, the bed is 1.95 m long by 1.60 m wide. Using the scale of 1 m = 10 squares, this results in a cutout rectangle of 19.5 × 16 squares. The other items of furniture were cut out in the same way. *Note*: Care needs to be taken with cupboards and cabinets, in order that they are placed so that the doors are able to swing open. Similarly, it is helpful to mark the way the bedroom door opens, again to ensure that it is not obstructed.

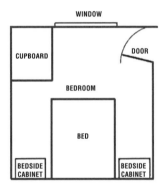

Appendix J

Measures of alcohol

Ethyl alcohol, chemical formula C_2H_5OH; an essence or spirit obtained by distillation.

Not everyone drinks alcohol, but whether you do or not, you will be aware of its effects. The classic symptoms of the drug are a feeling of well-being, associated with a slowing down of the thought processes and reduced ability to react quickly. Taken in excess, alcohol can damage the liver and cause problems of overweight.

So far, so bad! There is little doubt that, like cigarettes, if alcohol were invented today it would never be legalized!

If you or someone close to you does drink alcohol, it is sensible to know something about the alcoholic content of drinks and what sort of sensible limits are recommended by doctors.

Alcoholic content of drinks

Confusingly, there are two main ways of measuring how much alcohol there is in drink.

The old-fashioned measure of alcoholic strength is the degrees proof (e.g. 75° proof). In the UK, this is measured in the range between a minimum of zero and a maximum of 175 (in the US the range runs from 0 to 200). So water is 0° proof, and neat alcohol would be measured at 175° proof. This measure used to be applied to spirits and other drinks with a high alcohol content, but it is less commonly used these days.

Increasingly, bottles and cans of alcoholic drink are marked in terms of the percentage of alcohol in the drink (e.g. 8%). This may be stated on the bottle or can as '8% ABV', which means 8% alcohol by volume. Calculated in this way, neat alcohol would be measured at 100%. This measure has traditionally been applied to beers, cider, lager and other drinks with a relatively low alcohol content. However, most supermarkets and large retailers now use this method for spirits also.

The diagram below shows how to convert between these two measures.

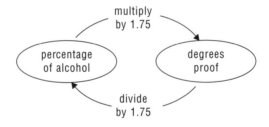

You might like to try the following exercise of converting between the two types of measure. And when you have completed the table, see if you can make some general comparisons between the strengths of different alcoholic drinks.

Drink	Degrees proof	Approximate % alcohol
White or red wine		14
Black Bush Irish Whiskey		40
Beauregard Napoleon Brandy		37.5
Safeway Vintage Port		20
Carlsberg Special (extra strength)	9.2	
Newcastle Brown Ale (strong ale)	7.5	
Woodpecker Cider (average cider)	5.5	
Tuborg Lager (ordinary lager)	3.7	

SOLUTION

Drink	Degrees proof	Approximate % alcohol
White or red wine	24.5	14
Black Bush Irish Whiskey	70	40
Beauregard Napoleon Brandy	65.6	37.5
Safeway Vintage Port	35	20
Carlsberg Special (extra strength)	9.2	5.3
Newcastle Brown Ale (strong ale)	7.5	4.3
Woodpecker Cider (average cider)	5.5	3.1
Tuborg Lager (ordinary lager)	3.7	2.1

There are a few interesting points to emerge from the table. Running your eye down the final column of the completed table, you can see that spirits like whiskey and brandy are nearly 20 times as strong, by volume, as ordinary lager. Also, a strong ale like Newcastle Brown contains twice as much alcohol as an ordinary lager. This means that drinking three pints of Newcastle Brown is equivalent to drinking six pints of Tuborg. Also, in terms of alcohol content, two pints of Carlsberg Special is roughly equivalent to five pints of Tuborg Lager.

Appendix K

Understanding barcodes

Up until the 1970s, supermarket goods were individually priced. Looking back, it is clear that this system had a number of drawbacks. Firstly, if the store decided to increase the price of, say, their baked beans, someone was required to collect all existing tins on the shelves, remove the existing labels and reprice each tin individually. Secondly, the system was open to abuse from dishonest customers who could switch price labels, substituting a cheaper one for a more expensive one before taking it through the check-out. Thirdly, each item had to be individually entered manually into the till at the check-out – a time-consuming task that was prone to error and abuse.

Barcodes changed all that. Instead of a price label being attached to each tin of beans or bag of muesli, etc., most items are manufactured with a barcode included on the packaging. A traditional one-dimensional barcode looks something like this.

Note that the information contained in a barcode is based on the widths of the spaces and not the lines. This type of barcode is still the most commonly used but more recently there has been a growth of two-dimensional barcodes that use dots, rectangles and other geometric patterns. Barcoded items are scanned electronically, a process which is almost error-free. So any information that is encoded in the barcode is transferred via the scanner into the computerized till. The beauty of the system is that the computer is able to log in much more information than simply the item's price. For example, each tin of beans that passes across the scanner is sold to a customer. This fact is automatically logged into the computer, so that the store has a running count of their stock at any given time. At the end of each day, they can then reorder new stocks of beans with some degree of precision. Precision in reordering is an important component in running a successful and competitive supermarket. Ordering insufficient tins of beans means the store may run out next day. Ordering too many results in a warehousing problem in storing the crates of surplus beans.

Sales over many months and years provide a valuable database of information, from which the store can predict seasonal patterns and so fine-tune their reorders. Also, price changes of an entire line can be entered as a single instruction on the store's computer, without staff having to reprice existing stock, item by item, on the shelves.

As you can see if you examine any barcode, the bars are written alongside a row of numbers. When the electronic scanner 'reads' the bars, the information being inputted is actually these numbers in coded form. There are different barcode systems; some have just 8 digits, while others have 13. Here is a 13-digit barcode for a 450 g tin of Heinz baked beans.

5 000157 004185

These 13 digits have been grouped rather oddly, with the first digit, the 5, on its own and the remaining 12 digits split into two groups of six. This is how the human eye sees the number, but the computer scanner groups them differently. In terms of what information the computer needs, the 13 digits split into four basic components, which are explained below.

50   The first 2 (sometimes 3) digits, are called the 'flag'. They indicate in which country the barcode was issued. *Note*: the flag does not tell you in which

country the item was produced, which may be different to the country of issue.

00157   The next five digits refer to the manufacturer; the number allocated to all Heinz products is 00157.

00418   The next five digits indicate the particular product. So, Heinz have allocated these five digits to refer to a 450 g tin of Heinz baked beans.

5   The final digit is known as a 'checksum'. Its purpose is to confirm that the digits recorded so far by the scanner are consistent and therefore likely to be correct. If the scanner should read the first 12 digits given above, followed by any digit other than 5, the computer will record an error (usually sounding a bleep) and the operator will know to rescan that item. The checksum is based on a formula applied to the previous 12 digits which should produce a single digit – in this case the number 5. The formula used for calculating this checksum is explained below.

The checksum

There are different ways of calculating checksums. This one, which applies to the popular 13-digit barcode, is calculated by completing the following stages.

Stage		Example
1	Number the first 12 digits from 1 to 12.	1 2 3 4 5 6 7 8 9 10 11 12 5 0 0 0 1 5 7 0 0 4 1 8
2	Add together all the odd-numbered digits.	$5 + 0 + 1 + 7 + 0 + 1 = 14$
3	Add together all the even-numbered digits.	$0 + 0 + 5 + 0 + 4 + 8 = 17$
4	Now add the result of Stage 2, the 14, to three times the result of Stage 3, the 17.	$14 + 3 \times 17 = 65$
5	Subtract the result of Stage 4 from the next bigger multiple of ten*, which in this case is 70.	$70 - 65 = 5$ This gives the checksum, 5, which becomes the 13th digit.

*Note: If Stage 4 had produced the result 82, you would subtract this number from 90; a result of 56 would have to be subtracted from 60, and so on. In the case where the formula produces a result ending in zero (say 60) then subtract it from itself, producing a checksum of 0 (60 – 60). The reason for setting up Stage 5 of the calculation in this form is to ensure a single-digit answer for the checksum.

You might like to explore this now for yourself. If you can't immediately lay your hands on any examples of 13-digit barcodes, then look at the start of this Appendix where one is reproduced. And here are two more to investigate.

Source	Barcode
Tesco Baked Beans 420g	5 018374 350930
Benecol Spread	7 640107 200335

a Can you work out the country of origin of these two items?

b Confirm that each checksum is correct.

c Suppose that a scanner misreads the first 12 digits of a barcode. How likely is it that the checksum would turn out to be correct for the incorrectly scanned number by chance alone?

SOLUTION

a As was mentioned in the text, the first two digits of these barcode numbers indicate the country in which the barcode was issued and not where the item was produced. The baked beans have a flag number of 50, which is the UK. Given the nature of the product it is likely that this tin of baked beans was in fact manufactured in the UK. The flag for the second item, 76, corresponds to Switzerland but again it is likely that this item was manufactured in the UK.

The checksums are calculated as follows.

Baked beans

$(5 + 1 + 3 + 4 + 5 + 9) + 3 \times (0 + 8 + 7 + 3 + 0 + 3)$

$= 27 + 3 \times 21$

$= 27 + 63 = 90.$

$90 - 90 = 0$ Check!

Benecol spread

$(7 + 4 + 1 + 7 + 0 + 3) + 3\times(6 + 0 + 0 + 2 + 0 + 3)$

$= 22 + 3\times11$

$= 22 + 33 = 55.$

$60 - 55 = 5$ Check!

b Finally, there is a one in ten chance of the checksum being accepted, even if the previous digits were scanned incorrectly. This is because there are ten possible digits available, and there is therefore a one in ten chance that the checksum digit scanned happened to match the other 12 by chance alone.

Appendix L

Junk mail and free offers

They say there is no such thing as a free lunch and the same principle probably applies to free offers, especially when they take the form of unsolicited junk mail. Many of the offers that arrive on your mat are packaged in the form of a game of chance, which you are invited to play. If you should be successful, and it's a fair bet that you will be, you qualify for their amazing free offer by being one of the very few lucky winners. You only need to complete the form and send away for big prizes. Further reading of the small print will no doubt reveal that things are not quite that simple.

A good example of this came through my door recently.

'Play this game and see how many mystery gifts you can claim!' the card read. The game card took the form of a 3 × 3 grid. Each cell in the grid was covered by a tear-off tab. The punter is asked to pull three tabs only. This revealed a number in each cell. If the numbers revealed added up to

6 .... *claim 1 gift*
7 .... *claim 2 gifts*
8 .... *claim 3 gifts*

My score came to 10, so I was clearly in luck. However, I couldn't resist tearing off the other six tabs (thereby making my game card void, but then that's life, eh!). The nine un-tabbed cells produced the following contents.

2	5	1
4	5	5
5	3	5

What this revealed was that the worst possible score I could get was $1 + 2 + 3 = 6$. In other words, I couldn't fail to win at least one prize. The second worst score I could get was $1 + 2 + 4 = 7$, which guaranteed two prizes. Any other combination guaranteed the maximum of three prizes. But just how many combinations are there altogether?

To calculate the number of possible selections, we go through each choice of cell in turn. There are 9 possible ways of choosing the first cell, 8 possible ways of choosing the second and 7 possible ways of choosing the third. Thus, the number of possible selections that I could have chosen is $9 \times 8 \times 7 = 504$. However, we need to be careful here, because each of these selections is the same as a number of other selections taken in a different order. In fact, for any selection of three things, there are six possible orderings.

If you aren't convinced of this, consider a selection of the letters ABC. The six different ways of ordering these are as follows.

ABC
ACB
BAC
BCA
CAB
CBA

So, we conclude that the figure of 504 is actually six times too large if you wish to count only the number of possible combinations, without taking account of order. This gives a final figure of $\frac{504}{6} = 84$ possible combinations from the game card.

So, it seems that, assuming the punters really do choose their tabs randomly, out of every 84 tries, the organizers could expect one person to win one prize, one to win two prizes and 82 to win all three. Put another way, something like $\frac{82}{84} = 0.976$, or 97.6% of punters will get the thrill of hitting the jackpot on this game.

Hmm! Maybe I wasn't quite as lucky as I thought!

Appendix M

Winning on the National Lottery

In case you have never bought a lottery ticket, here is how it works.

Numbers from 1 to 59 inclusive are printed out on a 'pay slip'.

You choose six numbers between 1 and 59. If at least three of the numbers you choose match any of the six main numbers drawn, you are a winner.

Prizes vary depending on how many numbers you can match. At the time of writing, the prizes are:

Winning selections	Expected prize
▶ Match 6 main numbers	Jackpot
▶ Match 5 main numbers + the bonus number	£50000
▶ Match 5 main numbers	£1000
▶ Match 4 main numbers	£100
▶ Match 3 main numbers Guaranteed	£25

Most countries run national lotteries. They provide a lot of fun and fantasy for the punters and, of course, are nice little earners for the government.

Lottery fever hit the UK in November 1994, with a much-hyped launch on TV, radio and the press. A lot of advice and information was offered to the great British public, most of which was total nonsense.

On the Jonathan Ross television show, a clairvoyant, named The Voxx, came up with a photo-fit of the winner:

> '...in her forties with strawberry blonde hair, possibly dyed, of Irish or Scottish background, has travelled extensively and has had a tough time in love but there is someone in her life at the moment. She also has a son.'

Well, that should narrow it down to a few hundred thousand people!

The *Sun* newspaper helpfully printed a giant dot, charged with lucky psychic energy. Readers were invited to touch the lucky spot, close their eyes and the numbers would just come to them by the sheer power of the 'Lottery spottery'. The *Sun* provided some evidence from the USA (where lucky spots were first devised) for this claim. Apparently, 'thousands of people said they only won because of their special power'. Wow!

Other papers offered yet more advice. For example, avoid so-called 'lucky' numbers like 7, 11 and 13, and avoid choosing numbers relating to birthdays or anniversaries. Can you think of any rational explanation for this?

So, who is to be believed, and is there a 'best strategy' for playing the lottery? Fortunately in answering these questions, mathematics can provide insights in to some of the parts that even lottery spottery cannot reach!

Let's try to sort out some of the fact from the fiction. Two key ideas will be explained below. The first explores the chance of winning – what sort of odds you are really up against. The second is to do with the way the numbers are chosen, both by the lottery 'random number generator' (the name for the machine that spits out the winning numbers) and by the paying, playing punters.

What are my chances?

Roughly half of the money paid into the lottery is given back in prizes. So, taking a very long-term view, if you bought, say, £1000 worth of lottery tickets over your lifetime, you could expect, on average, to lose about £500. The reality is that you will almost certainly not win one of the monster prizes, but then again you just might. A typical jackpot may be anything from £2 million to £80 million, depending on whether there are un-won jackpots rolled over from previous weeks. According to the promoters, the odds against winning the jackpot (or a share of it, if other players chose the same numbers as you) are about 45 million to one. By the way, if you are interested in how this figure of 45 million is calculated, it is explained at the end of this appendix. So for this prize, you would expect, on average, to have to lay out £45 million to win back as little as £2 million. At the other end of the winnings scale, there is one chance in 97 of winning a £25 guaranteed prize – i.e. you would expect, on average to lay out £97 to win back £25.

In terms of return on your investment, this is pretty thin gruel, whichever way you serve it up. But then again, what keeps most of us losing money on such foolishness is that we might just win that big one!

How are the numbers chosen?

As with Bingo, a key principle of the lottery is equal likelihood – i.e. the device for choosing the numbers is designed so that each number has an equal chance of coming up. The only thing that would prevent that from happening is if the random number generator was programmed to generate numbers in a different way. But then that would be cheating.

Numbers which have an equal chance of coming up are known as *random numbers* (for example, tossing dice, coins, and so on). So, since lottery numbers are chosen at random, no number or combination of numbers is more or less likely to come up than any other. For example, the selection 1, 2, 3, 4, 5, 6 is just as likely (well, just as *unlikely* would be more appropriate) as a mixed bag of numbers like 32, 6, 18, 41, 9, 15.

Clearly, then, you have no control over whether or not your numbers win. But what about sharing your winnings with others? Here you *can* exercise your skill and judgement by anticipating what numbers other punters are likely to choose. Basically, if you come up with a winning combination, you will share it with fewer people if you pick numbers that others are less likely to pick. So this is where a bit of mind reading comes in. A simple example may make this clearer.

A simplified lottery example

Twenty punters, paying £1 each to play, choose a number in the range 1 to 6. A die is tossed and the £10 winnings[1] are shared among those who chose the winning number.

The punters' choices are shown below.

Selection	1	2	3	4	5	6
No. of people	2	5	4	5	3	1

In other words, two punters chose '1', five chose '2', four chose '3', and so on. Now, notice that if the die shows up '2' or '4', the winnings have to be shared among five winners, so each of these winning punters gets $\frac{£10}{5}$ = £2. But it is just as likely that the winning number turns out to be '6', in which case the lucky winner scoops the lot. In fact, '6' would be a good choice here, because (due to negative experiences playing board games) people tend to avoid this number, in the mistaken belief that it is less likely to come up than any other.

[1] 'Hey, where did the other £10 go?'

To the lottery organizers, of course. There are so many expenses; they have huge advertising and administrative costs. Then they have to buy the die, and that doesn't come cheap. And, of course, they need to train their staff to toss it and see fair play all round.

So, a good overall strategy is to avoid numbers that other people are likely to choose and to go for numbers that they are unlikely to choose.

As far as the National Lottery is concerned, that means:

A good overall strategy

Avoid   people's 'lucky' numbers, like 3, 7, 11.

Avoid   numbers linked to birthdays or anniversaries.

            In other words avoid all numbers of 31 or below (days in a month) and particularly avoid numbers of 12 or below (months in a year).

Go for   numbers above 31.

Go for   numbers in a sequence (many people mistakenly think that these are less likely than numbers which 'look random').

Finally, note that these strategies are only successful if we have successfully predicted how the other punters will choose their numbers (i.e. that they will tend to go for numbers lower than 31, avoid numbers in sequence, and so on). In fact, all the indications are that this is indeed what people do. Evidence for this emerged from the very first UK national lottery in November 1994, which produced the following winning numbers:

3, 5, 14, 22, 30 and 44 (10 was the bonus number)

To the great disappointment of the organizers, Camelot, the jackpot had to be scaled down from an estimated £7 million to £5.8 million as the number of small-scale winners became known. Camelot staff had not expected there would be as many as the one million players who would pick up the guaranteed £10 which was, at that time, the prize pay-out for three correct numbers. The reason, it seems, was that most of the players did pick numbers relating to birthdays and, as can be seen, five of the six winning numbers were below 31.

So, the best strategy at the time of writing, is to choose numbers above 31 and opt for strings, rather than avoid them, as the naive players will. Of course, at a certain point, when enough people have read this book, it may be the case that the informed punters will outnumber the naive ones. When that happy state arrives, you may feel it appropriate to change your strategy accordingly. Me? Well, if this knowledge is brought about by my book selling more than 5 million copies, I certainly won't need to waste my time or money on a lottery!

How is the figure of 45 million to one calculated?

It is correctly claimed by the lottery organizers that if you choose six numbers at random from a list of 59, the odds are about 45 million to one against your selecting the six winning numbers. The calculation is explained below based on an initial incorrect assumption, but I will sort that out at the end!

Start by choosing the first number. There are 59 to choose from so there are clearly 59 choices. Choosing the second number means choosing from the 58 remaining numbers. So, for each of the 59 first choices there are 58 second choices. In other words, there are 59 × 58 ways of choosing the first two numbers. Similarly, there are 59 × 58 × 57 ways of choosing the first three numbers, and so on. Following the same line of argument, it would seem that the number of ways of choosing the first six numbers correctly are:

$$59 \times 58 \times 57 \times 56 \times 55 \times 54$$

Unfortunately, most basic calculators are unable to perform this calculation because it produces a result too big for the calculator to display. So will you take my word for it that this is the correct answer? Well, you shouldn't! As I suggested earlier, there is an error in the reasoning here, which I will now correct.

The error is that I have included each combination of numbers many times, as a result of which this answer is too large. To convince you of this, imagine that the first set of six numbers you chose was: 1, 2, 3, 4, 5 and 6.

This same combination of numbers could also crop up as 1, 3, 2, 4, 5, 6 or 6, 5, 4, 3, 2, 1 or any ordering you can think of. But just how many orderings are there? This question was explored in Appendix L: Junk mail and free offers. There we ordered three things and found that there were six possible orderings. The number of ways of ordering six things is more complicated and can be calculated as follows.

There are six ways of ordering the first number, five ways of ordering the second, four ways of ordering the third, and so on.

So, the number of ways of ordering six things = 6 × 5 × 4 × 3 × 2 × 1 = 720.

And now, back to the plot. This last discussion suggests that each combination of numbers contained in the calculation

$$59 \times 58 \times 57 \times 56 \times 55 \times 54$$

is actually included 720 times.

So, the number of separate combinations is

$$\frac{59 \times 58 \times 57 \times 56 \times 55 \times 54}{720}.$$

Now we can punch this out on the calculator and hope we get the answer 45 million. One snag is that, if you calculate the top line first before dividing by 720, you will almost certainly cause the calculator to overflow. A sneaky way out of this is to divide by the 720 sooner rather than later in the calculation. For example, I pressed the following:

59 ÷ 720 × 58 × 57 × 56 × 55 × 54 =

On my calculator, this produced the answer 45 057 474, which isn't all that far away from the result we were hoping for of 45 million.

Appendix N

Safe travel

Is it safer to travel by car or by motorcycle?
Most people would say that car travel is safer, but is this really true?

In 2014, there were 1713 fatalities resulting from road accidents in Great Britain, of which 797 were car occupants, 446 pedestrians, 339 motor cyclists and 113 pedal cyclists.

Note: the sources for this information are available online at www.gov.uk. Search for 'rail trends' and 'road fatalities', respectively.

So at first glance it would seem that motorcycle travel is safer, since more car occupants are killed than passengers on motorcycles. However, the comparison is not a fair one because we are not comparing like with like. An important complicating factor is that many more people travel many more kilometres in a car than by motorcycle, so the car occupants are exposed to risk on more occasions.

To make a fair comparison, we need to take account of the average distances travelled by each mode of transport and use these to calculate the fatality rates. These can then be compared directly. So here goes.

The fairest figure to use here is the number of passenger kilometres for both car and motorcycle users. As the name implies, one passenger kilometre is recorded when one passenger travels one kilometre. If five passengers each travel 10 kilometres, a total of 50 passenger kilometres will be recorded.

Typical annual car occupant use   664 billion passenger km

Typical annual motorcycle   4.8 billion passenger km
   passenger use

To calculate the accident rate per billion passenger kilometres, divide the number of deaths in a year by the number of billion passenger kilometres travelled. The result is fatality rate per billion passenger kilometres.

Car user death rate $= \frac{797}{664} = 1.2$

Motorcycle user death rate $= \frac{339}{4.8} = 70$

Using this comparison, it seems that motorcycle travel is roughly sixty times as dangerous as travelling in a car.

It is interesting to look at other forms of transport, based on this comparison of the number of deaths per billion passenger kilometres. Typical annual figures for Great Britain are as follows.

Mode	Rate per billion passenger kilometres
Air	0.01
Water	0.8
Rail	0.0
Bus or coach	0.2
Car	1.2
Van	0.3
Motorcycles	70.0
Pedal cyclists	22.0
Foot	24.0

Source: Department of Transport

So, according to these figures, air and rail travel are the safest forms of transport and motorcycling is very much the most dangerous. However, it should be stressed that all measures have their drawbacks and this is just one possible measure. A weakness in the measure used here is that it favours modes of travel which are fast over the slower methods. Thus, air travel can allow you to cover, say, a hundred kilometres in just a few minutes, whereas you would take days on foot to cover this sort of distance. As a result, for a given number of passenger kilometres, the exposure to risk on foot is much greater merely due to the fact that there is a longer period of time during which an accident can happen. However, provided this aspect is borne in mind, using the number of deaths per billion passenger kilometres is probably the fairest comparison available.

Taking it further

Websites and organizations

Association of Teachers of Mathematics (UK)
atm.org.uk

Mathematical Association (UK)
m-a.org.uk

Math in Pictures is one of the author's personal sites, which includes many resources to support learning, including over 75 short mathematical videos.
mathinpictures.co.uk

Maths is Fun provides many well-designed resources for mathematics learners.
mathsisfun.com

National Council of Teachers of Mathematics (USA)
nctm.org

National Library of Virtual Manipulatives, containing a useful collection of visual resources from Utah State University.
nlvm.usu.ed

NRICH, which is linked to Cambridge University, offers a wide range of excellent resources to support mathematical earning.
nrich.maths.org

UK National Statistics online
statistics.gov.uk

Reading list

Barrow, John D., *Pi in the Sky: Counting, Thinking and Being* (London: Penguin, 1993). An exploration of where maths comes from and how it is performed.

Blastland, M. and Dilnot, A., *The Tiger That Isn't: Seeing Through a World of Numbers* (London: Profile Books, 2008). Mathematics in the real world – both funny and highly informative.

Eastaway, Rob and Wells, David, *Mindbenders and Brainteasers* (London: Robson Books, 2005). A collection of 100 puzzles and conundrums, old and new.

Eastaway, Rob and Wyndham, Jeremy, *How Long is a Piece of String?* (London: Robson Books, 2002). Examples of mathematics in everyday life.

Eastaway, Rob and Wyndham, Jeremy, *Why Do Buses Come in Threes?* (London: Robson Books, 1998). Practical uses for various mathematical topics, including probability, Venn diagrams and prime numbers.

Flannery, Sarah, *In Code: A Mathematical Journey* (London: Profile Books, 2000). A collection of problems with solutions and explanations, based on the author's experiences of growing up in a mathematical home.

Goldacre, Ben, *Bad Science* (London: Harper Perennial, 2009). One of the best books I've come across for debunking myths about medicine, beauty products, brain-training, homeopathy and much more. This book should be on the National Curriculum.

Graham, Alan, *Improve Your Maths (Teach Yourself Your Evening Class)* (London: Hodder Education, 2008). The combination of written and DVD formats should give you that extra leg up which you may not get from a book alone.

Graham, Alan, *Mathematics Made Easy: Flash* (London: Hodder Education, 2011). In just 96 pages, the reader will learn all the basics, from addition and subtraction to fractions and decimals. Ideal for the busy, the time-pressured or the merely curious.

Graham, Alan, *Statistics Made Easy: Flash* (London: Hodder Education, 2011). A short, simple and to-the-point guide to statistics that explains why we do the things we do in this subject.

Graham, Alan, *The Sum of You* (London: Hodder Education, 2011). In this groundbreaking book, Alan Graham argues that there are six basic different personality traits present in all of us, to a greater or lesser degree. He uses these personalities as starting points for exploring the big ideas in mathematics.

Graham, Alan, *Statistics: An Introduction, Teach Yourself* (London: Hodder Education, 2017). A straightforward and accessible account of the big ideas of statistics with a minimum of hard mathematics.

Haighton, June et al., *Maths: The Basic Skills* (Cheltenham: Nelson Thornes, 2004). A traditional textbook on basic mathematics based on the Adult Numeracy Core Curriculum.

Huntley, H.E., *The Divine Proportion, Study in Mathematical Beauty* (New York: Dover, 1970). Applications in art and nature of the 'Golden Ratio'.

Ifrah, Georges, *The Universal History of Numbers* (London: The Harvill Press, 1998). A detailed book (translated from French) about the history of numbers and counting from prehistory to the age of the computer.

Paulos, John Allen, *Innumeracy – Mathematical Illiteracy and its Consequences* (London: Penguin, 1990). Real-world examples of innumeracy, including stock scams, risk perception and election statistics.

Pólya, G., *How to Solve It* (London: Penguin, 1990). A classic text on mathematical problem solving that is well known around the world.

Potter, Lawrence, *Mathematics Minus Fear* (London: Marion Boyars Publishers Ltd, 2006). A romp through school mathematics that takes in puzzles and gambling.

Singh, Simon, *The Code Book* (London: Fourth Estate, 2000). A history of codes and ciphers and their modern applications in electronic security.

Singh, Simon, *Fermat's Last Theorem* (London: Fourth Estate, 1998). An account of Andrew Wiles' proof of Fermat's Last

Theorem, but also outlining some problems that have interested mathematicians over many centuries.

Stewart, Ian, *From Here to Infinity* (Oxford: Oxford University Press, 1996). An introduction to how mathematical ideas are developing today.

Stewart, Ian, *Does God Play Dice?* (London: Penguin, 1997). An introduction to the theory and practice of chaos and fractals.

Stewart, Ian, *Letters to a Young Mathematician* (New York: Basic Books, 2006). What the author wishes he had known about mathematics when he was a student.

Stewart, Ian, *Taming the Infinite* (London: Quercus Publishing plc., 2008). A clear and interesting account of the history of mathematics, aimed at the non-technical reader.

Index